Reason to Be Happy

Reason to Be Happy

Why logical thinking is the key to a better life

Kaushik Basu

torva

TRANSWORLD PUBLISHERS

Penguin Random House, One Embassy Gardens,
8 Viaduct Gardens, London SW11 7BW
www.penguin.co.uk

Transworld is part of the Penguin Random House group of companies
whose addresses can be found at global.penguinrandomhouse.com

First published in Great Britain in 2024 by Torva
an imprint of Transworld Publishers

A CIP catalogue record for this book
is available from the British Library.

ISBN 9781911709282

Typeset in Minion Pro by Falcon Oast Graphic Art Ltd
Text design by Couper Street Type Co.
Printed and bound in Great Britain by Clays Ltd, Elcograf S.p.A.

The authorized representative in the EEA is Penguin Random House Ireland,
Morrison Chambers, 32 Nassau Street, Dublin D02 YH68.

Penguin Random House is committed to a sustainable
future for our business, our readers and our planet. This book is
made from Forest Stewardship Council® certified paper.

For my students,
in the hope of a better world

Contents

Preface

THIS BOOK HAS LIVED with me for a long time. For years, I have been tucking away ideas in my head and on scraps of paper about how to navigate careers and life, both the highs and the lows. In writing the book I have drawn not just on academic writings on game theory, economics and philosophy but also on my experience, from my childhood in India to my working life as an economist and as a policymaker, trying to tackle real-world problems. Much of what I have learned came from my interactions with others – students in classrooms, embattled politicians and challenged corporate leaders whom I met during my years as the chief economist of the World Bank in Washington, and as chief economic adviser to the Government of India. In working on this book, I have built up a lot of debt, as much to people as to places. I have led a rather nomadic life, spanning many countries, myriad cities and towns.

It was not meant to be that way. As far back as my memory goes, the plan seemed to have been that I would become a lawyer and take over my father's flourishing law firm, P. C. Ghose & Co. in Kolkata (Calcutta, at that time). I loved Kolkata with its intellectual bustle, coffee houses and bookshops. It seemed an agreeable plan. I would get married, have children and live my life at 29/1 Ballygunge Circular Road, the sprawling home to which I moved with my parents as a child. I would work on the first floor (second, in Americanese) of the horizontally stretched office building at 8/2 Hastings Street, with its four rooms and French windows. The building had a long balcony, where clerks and clients, weighed down by worries about stolen property, contested wills and unpaid rents, would be waiting to meet the lawyer, while watching the scene below: pedestrians in jaded suits, dhotis and saris, a cacophony of horns, each driver urging the one in front to move, unmindful of the fact that the driver in front had another driver in front, who had another driver in front.

It was in London, as a student at the London School of Economics, that I changed my career plan. London in those days was a hub of progressive movements and new ideas. Sitting in the crowded lecture halls, listening to some of the world's finest minds ponder economics, philosophy and politics, I wondered at the way they unearthed patterns in the world that others had missed. There I met Amartya Sen, who, as professor and later my PhD adviser, was the most important reason for my career change, and it was his lectures on social choice theory and investment planning which got me interested in economics and philosophy.

The Delhi School of Economics, where I began my research and teaching career, was an amazing place in the 1980s and 1990s. In the spirit of Paris's famous Left Bank, some of the finest minds of Delhi University were more likely to be found in the coffee house on the premises of the School than their offices. The next big move happened with an unexpected professorship offer from Cornell University. It was a move made with some trepidation. My wife, Alaka, and I had been comfortably settled in Delhi for seventeen years, and we were not looking to move. In retrospect, I was lucky I did. I was beginning to drown in administrative work. Cornell gave me the break I needed.

The move to Ithaca was wonderful. I quickly found a home in a surprisingly multidisciplinary intellectual hub, which has contributed to this book. It began with a mild embarrassment. Before I knew I would move to Cornell, I had written a review of the book *Harold Laski: A Life on the Left* for an Indian newspaper. Having no idea who the authors were, I paid no attention to how I worded my criticisms. Within days of moving into my new office in Uris Hall, I got a phone call from one of the authors of the book, Isaac Kramnick. He said he had read my review and, much to my alarm, added that he was three or four buildings away from where I was sitting, in the Department of Government. We agreed to meet for lunch. Isaac Kramnick turned out to be one of the most charming, irreverent and warm human beings I have met. We quickly became friends, and through him my circle of friends beyond economics grew rapidly. Later, Alaka and I would be part of a group of individuals who straddled multiple disciplines and met regularly. This included Mary and Peter Katzenstein, Glenn Altschuler, Elizabeth and Hunter Rawlings, Vivian and Gary Fields, Charla and Erik Thorbecke and several others. For my interest in politics, philosophy and development economics, I owe a lot to this enormously talented and equally talkative group.

Despite this broad interest, in my research I remained actively involved in fairly mainstream economics. Although the ideas in this book had been building up for a long time, there is a specific date which triggered the writing of the book. I was invited to give a talk to Cornell's Sage School of Philosophy on 27 March 2019. It was meant to be an informal talk, an occasion to let the bees out of one's bonnet. The discussion, with comments from Avi Appel, Tad Brennan, Nicole Hassoun, Rachana Kamtekar and Theodore Korzukhin, got the ball rolling. Before I knew it, I was scribbling notes and systematically jotting down passing thoughts.

Another important landmark was an invitation from Geoffrey Sayre-McCord to address the annual meeting of the Philosophy, Politics and Economics (PPE) Society in New Orleans in February 2022. This was an introduction to a world of which I was subliminally aware but with which I had had no interaction. It turned out to be an opportunity to talk to a wide range of philosophers and political scientists (when not exploring New Orleans) and to try out some of my thoughts on moral responsibility on a large, interdisciplinary audience.

The perfect break to work on the manuscript at a stretch happened when I got an invitation to spend the summer of 2022 at the Bucerius Law School in Hamburg. The nudge to bring the book to closure came from the many discussions I had with my friend Hans-Bernd Schaefer at the Law School, and the free-flowing conversation I had with Hartmut Kliemt and Marlies Ahlert over two days in pastoral Lower Saxony. Soon I had a draft, however patchy, of the full book.

The final editorial work was done during my sabbatical, in spring 2023, while visiting the London School of Economics. London was my first encounter with the world outside of India when I arrived here in 1972 for my graduate education. It was with a feeling of nostalgia that I sat here in 2023, working, once again, like a student, putting final touches to the manuscript.

My indebtedness has piled up over this long haul. First of all, I have drawn extensively on the works of economists, game theorists and philosophers. I have tried to be careful in citing the relevant works, but there are bound to be omissions. I am particularly conscious of the vast literature in philosophy and my own eccentric and patchy reading. While working on the book, I have, every now and then, discovered writings of relevance with which I should have been familiar. By induction, I assume that there are more readings I have missed. My apologies for all such omissions. I hope to make amends in future editions. I have also drawn on some of my own published works, mindful of the fact that the readership for this book will likely be different from my usual readers.

For comments on the material covered in the book and discussions on related topics, I would like to thank, in addition to the names already mentioned, Karna Basu, Larry Blume, Stephen Coate, Chris Cotton, Avinash Dixit, Julia Markowitz, Ajit Mishra, Michael Moehler, Puran Mongia, Ryan Muldoon, Derk Pereboom, Sudipta Sarangi, Neelam Sethi and Himanjali Shankar.

I would also like to take this opportunity to thank my teaching and research assistants and recent students Valeria Bodishtianu, Aviv Caspi, Jacob Fisher, Meir Friedenberg, Chenyang Li, Fikri Pitsuwan, Haokun Sun, Saloni Vadeyar, Pengfei Zhang and Zihui Zhao, for related discussions, comments and editorial help. In the final phase of this work, I benefited greatly from the comments and counsel of Andrew Gordon and Alex Christofi.

Finally, I am grateful to my wife, Alaka, and children, Karna and Diksha, and their partners, Shabnam and Mikey, for agreements, disagreements, discussion and laughter. Alaka read through the entire manuscript, commenting in detail on the content, logic and language. Her speed of reading always makes me wonder if she actually reads. Her capacity to spot 'errors', as she calls them, on every page, confirms that she does.

1

Reasoning, Happiness and Success

MUCH OF OUR LIFE is spent seeking happiness and contentment, and navigating hurdles and challenges. One of the most powerful and underutilized tools for this journey is one that we all possess; namely, the ability to reason. Hence, the two meanings of the book's title. Trying to make sense of the world around us and defining a meaningful way of life is what the philosophers of antiquity were concerned with. This book is a revisit to these ancient topics, but with the ammunition of modern disciplines, notably economics and game theory. Along the way, and in addition, the book explores the philosophical cracks and paradoxes that underlie these disciplines, and the moral quandaries to which they give rise, and thereby hopes to engage its readers in the quest for knowledge and understanding.

The book is organized as an inverted pyramid. The initial focus is on the individual, and on rational choice and reasoning in everyday life, ranging from office politics to personal dilemmas. Thereafter, the book moves on to consider the well-being of the collective and the moral responsibility of groups. From the individual and the group, the final chapter turns to our largest concern, namely, the globe, to discuss how we might tackle some of the problems facing our careening world. Can we as individuals make the world a better place, and if so, how? To begin to find an answer, I'd like to start by showing how game theory can save lives.

'If I were you'

A few weeks after my arrival in London in 1972 as a student at the London School of Economics, I was walking through Brunswick Square when a kid threw a water balloon at me. It grazed my shoulder and did no damage. Callow, new to London, and in fact outside my native India for the first time, I hesitated for a moment, wondering if it was worth picking a fight. I decided there was no reason to do so, and continued walking.

Just then, a well-built man strode up to me and, making no effort to hide his disdain (more for me than the boy), said, 'If I were you, young man, I would have thrashed that boy.' My immediate reaction was, 'No, you would not, because, as you just saw, I did not.' Needless to say, given his predisposition to use physical force, I did not say this out loud.

If I were you is an important part of moral reasoning, from Kantian ethics to welfare economics and game theory. Yet, it is widely misunderstood. If it means literally what it says, then once 'I' see how 'you' behave, there is no further reason to speculate on what I would do if I were you. If, on the other hand, 'if I were you' meant 'if I were *partly like* you', a host of interesting questions would arise. There would be a lot of ambiguity, too, because there are many possible ways in which I could be like you.

Game theory involves a lot of reasoning of this kind. It's not enough to be clever; you must be able to put yourself in the shoes of the other clever person and think of what she might do – which, in turn, is of course based on what she thinks you might do.

Given that Brunswick Square is close to the University of London, the man may well have been a professor. He may also have been good at research. However, his emotions got the better of him here. This is a common human propensity and the source of many errors of judgement, and also a large part of the motivation behind this book.

This book is meant to show both why we have reason to be happy, and why we must reason to be happy. To make my case, I will draw on the kind of reasoning used in game theory. Game theory is the art of deductive reasoning in social situations. For that reason, it is of value in war and diplomacy, in developing corporate strategy, and even in our day-to-day interpersonal relationships.

On Tuesday, 16 October 1962, a little before 9 a.m., President Kennedy was informed by his National Security Advisor, McGeorge Bundy, that a US U-2 reconnaissance plane had discovered that the Soviets had placed nuclear-armed ballistic missiles in Cuba, which could hit American cities within minutes. Kennedy's immediate problem was to decide what he and the United States should do. That decision was critically dependent on what he thought the Soviet premier, Nikita Khrushchev, would do in response to what he did. And he was no doubt aware that what Khrushchev would do would depend on what Khrushchev thought Kennedy would do in response to what Khrushchev did. This was a classic game-theoretic problem, involving reasoning of the kind used in playing chess or bridge. The difference here was that the stakes were life and death.

The following thirteen days would go down in history as among the most dangerous humanity has ever witnessed. The world watched this chilling 'game' by the minute. Nearly 180 bombers, carrying nuclear weapons, were instructed to be continuously airborne, flying up to the Soviet border and back. The message was clear: if there were an attack on the United States, the United States might perish but so would Russia. This was the so-called 'second-strike strategy', which is meant to deter anyone contemplating the first strike. Once some countries had nuclear capability, for another country contemplating producing nuclear bombs, the second-strike capability was thought of as essential. To have the bomb and not the second-strike capacity is to invite an attack. The open display of second-strike capacity by the United States worked. Khrushchev eventually backed down, and the world was saved.

What received little or no attention, mainly because they happened in secrecy, were the long hours of discussion of strategy and counter-strategy that took place in the White House. In fact, it took Kennedy six nerve-wracking days to respond. After long deliberation, the US president decided to go public on 22 October, informing the American people of the peril they were facing, and writing to Khrushchev delineating how the United States planned to act.[1] The world owes a lot to those six days of master strategizing, led by Kennedy, that enabled a quiet wind-down from the brink of world war and destruction. Khrushchev also deserves credit for opting to lose face rather than the world.

What helped Kennedy play this game of war so well was his Bay of Pigs debacle of more than a year earlier, in which a US-engineered military operation by Cuban exiles had been foiled by Cuba's armed forces with embarrassing speed. He realized that the attempt to overthrow the communist government of Cuba had failed not because the US did not have more guns and weapons but because he did not allow enough time to deliberate and strategize. In the words of May and Zelikow (1997, p. 14), the Bay of Pigs failure made him realize 'he had not only listened to too few advisers but that he had given the issues too little time'. One consequence of this was that Thomas Schelling, who would later win the Nobel Prize in Economics for his pioneering work on game theory, was asked to write a paper on nuclear strategy, which he did and submitted on 5 July 1961. This paper, according to McGeorge Bundy, made a deep impression on the President. That, in turn, played a huge role in steering the nation through the Cuban Missile Crisis the following year.

As I write this book, President Biden is engaged in responding to Vladimir Putin's invasion of Ukraine. For each support measure he contemplates, he has to think about what Putin will do in response. This is fraught because there are doubts about Putin's rationality. The CNN documentary *Inside the Mind of Vladimir Putin*, presented by Fareed Zakaria, is fascinating to watch. Its only problem is that, at the end of watching it, you will have no idea what is going on inside the mind of Vladimir Putin. Even though game theory is one of the most exciting disciplines that emerged in the twentieth century, it has its blind spots, where we have to use intuition, psychology, politics and philosophy. And we must be reconciled to the fact that not every problem has a solution. By combining insights from these different approaches, we can only hope to increase the likelihood of success.

We have evolved over millennia to intuit what other people might be thinking, and that provides some of the foundations of game theory. But, as a coherent, self-contained discipline, game theory is surprisingly young. Émile Borel, the celebrated French mathematician and, later, politician, produced some of the most important works on game theory in 1921. However, as a method of analysis, influencing a staggering range of subjects, from economics, politics and psychology to evolutionary biology, computer science and philosophy, game theory burst upon us in the middle of the twentieth century – like Philip Larkin's first epicurean encounter, somewhere between the end of the *Chatterley* ban and the Beatles' first LP. It is perhaps no accident that the arrival of game theory coincided roughly with the invention of the atom bomb, as well as the emergence of global corporations competing with one another and with governments in novel ways. By providing tools of analysis for war, diplomacy and corporate strategy, it shaped some of the dimensions of the modern world.

The following pages will use the story of game theory to illustrate how we can meet life's challenges. However, not everything in the book is meant to help directly to solve a specific problem. This book, taken in a few pages at a time, should do for the mind what jogging does for the body. We go jogging not because it generates output or income, but to enhance our physical well-being so that we can be more effective when we do everything else. Likewise, logic and game theory can help train our minds, so that when there's something we need, we can get it more effectively. Some readers may wish to treat reading a few pages of the book every day as a substitute for the daily sudoku or crossword.

In the process of this exercise, the book takes the reader to some important conundrums and paradoxes. Like jogging, these can be enjoyable in themselves and therapeutic for the mind. As the early Greek philosophers, especially the Stoics, were aware, philosophy is not just an intellectual quest but a way of life. And if, in the process, you manage to solve one of the unsolved paradoxes, even if you get no happiness from it, you could go down in history as a philosopher, if that is any consolation.

Bertrand Russell's chicken

The caveat, 'if that is any consolation', is necessary. Take, for example, the concern of Jean-Baptiste le Rond d'Alembert's foster mother. D'Alembert was abandoned by his biological mother on the steps of the Saint-Jean-le-Rond church in Paris, within days of his birth on 16 November 1717. He acquired his name from the church and grew up initially in an orphanage and then with a foster mother. Miraculously, despite a difficult start in life, he became one of the world's greatest thinkers, making vital contributions to mathematics, philosophy, music theory and physics. Nevertheless, his foster mother remained disappointed that he never seemed to do any useful work. When d'Alembert told her about one of his discoveries, she famously responded, 'You will never be anything but a philosopher – and what is that but an ass who plagues himself all his life, that he may be talked about after he is dead.'[2]

It is true that philosophy and all pursuits of the mind have an element of escapism. There is no harm in this. They help one achieve equanimity of the mind (*ataraxia*), promoted by sceptics like Pyrrho of Elis and Stoics like Zeno of Citium, Diogenes of Babylon and Epictetus. At the same time, the quest for truth and science can, on some occasions, lead to intellectual breakthroughs that vastly expand our understanding of the universe and provide ingredients for creating a better world.

It is at times said that philosophy as a discipline began on 28 May 585 BCE. That day, not for the first time, there was a solar eclipse. The difference this time was that Thales of Miletus had predicted it would take place. Thales was a master of geometry and had proved a beautiful theorem about circles and right-angled triangles. The prediction of the eclipse, with hardly any instruments for watching the sky up close, was the outcome of deep contemplation and speculation. It was a major intellectual breakthrough for humankind, the outcome of months and years of tracking the movements of planets and stars, and inductive and deductive reasoning.

Inductive reasoning involves looking at patterns in nature and, from those patterns, drawing conclusions about the future. Most of us believe that the sun will rise tomorrow because we have seen it rise, with great regularity, in the past. This is inductive reasoning.

Deduction, on the other hand, entails reaching conclusions from premises based on pure logic. The truth is already contained in the premises. In a right-angled triangle, the area of the square of the hypotenuse is equal to the sum of the areas of the squares on the two other sides. We don't need to collect triangles from around the world, do careful measurement and then reach the conclusion. The conclusion follows from the definitions of concepts like triangle, right angle and square. It is so transparent that, in principle, anyone can see it. It is another matter that, before Pythagoras, no one saw it, and even now, over 2,500 years after Pythagoras, school students are often reduced to tears trying to see it too.[3]

What got me hooked on philosophy early in life was a delightful observation on inductive reasoning by Bertrand Russell in his slim monograph *The Problems of Philosophy* (p. 98): 'The man who has fed the chicken every day throughout its life at last wrings its neck instead, showing that more refined views as to the uniformity of nature would have been useful to the chicken.' This quote perfectly captures the pitfalls of induction. It prods us towards scepticism. It is a plea for reason. I believe there is a lot to learn from it.

My only disagreement with Russell is with the prescriptive part. I would modify it slightly, to: The man who has fed the chicken every day throughout its life at last wrings its neck instead, thereby showing humans the benefit of living life like the chicken.

Arguably, there is nothing the chicken could do to avoid ending up on the dinner table. To fret about something over which you have no control is to bring unnecessary distress upon yourself. In any situation, it is worth pondering the 'actions' or 'strategies' which are available, or 'feasible', in order to choose well and maximize money, power, fame, or whatever it is that you seek. However, we human beings spend a disproportionate amount of time fretting about matters over which we have no control. If there is nothing the chicken can do about its fate, getting angry with the man does nothing but take away some value from the chicken's own life. By saying we should live life like the chicken, I do not mean we should delude ourselves into believing that the future will be like the past. But where we know that there is nothing we can do to change the course of our lives, we should try to live life *as though* the future will be as good as the past.

Russell's thought experiment draws attention to the problem with induction. The fact that the sun has risen every day is no guarantee that it will rise tomorrow. Of course, we live by induction. We observe patterns in nature or, more accurately, observe nature and create patterns in our heads. There is, however, no objective reason for this. If someone says, despite all the scientific evidence, she does not believe vaccination protects her against COVID-19, there is no compelling reason to dismiss her claim. The difference between what we call science and superstition is not as great as we imagine. Honesty demands that we be sceptics.

The French photographer Eric Valli worked extensively among the Gurung tribespeople of Nepal, who climb dangerously high trees to gather honey. He once asked one of them if they ever fall down from their high perch. The man's deadpan response was: 'Yes, you fall when your life is over.'[4] I am urging the reader not to be totally dismissive of the Gurung point of view.

Induction is important, but we need to be aware of its shortcomings for discovering causality. Suppose researchers come to a community to test the effect of injecting the inhabitants with a green chemical to improve their memory.[5] They test a large number and find that it improves memory greatly in each and every case and has no negative side effects. If the study is properly done, it is likely to be treated as an important finding and will get published in a major scientific journal. Now consider a woman called Eve who lives in this town, was not subjected to the test and is keen to enhance her memory. Can she, on the basis of this scientific study, deduce that her memory will improve if she takes this injection? Believers in induction will confidently say yes. To see why their confidence is suspect, let me fill in some of the details of the above scientific experiment.

Suppose the community happens to be inhabited by a large number of snakes, frogs, rats and bats, and just one human being, Eve. So what happened was that a large sample of creatures (but no humans) were administered the green injection and got the result described above. Clearly, Eve will have reason to hesitate over whether to take this green injection. It is true that it worked in the case of snakes, frogs, rats and bats. It's also true that she is part of the population from which the scientific sample was drawn, but this would not be of much consolation.

In fact, if news comes from a neighbouring town, where all living creatures happen to be humans, that the experiment was done on only one inhabitant, who was given the injection, and not only did it not enhance her memory, but she got a throbbing headache, then we will all have sympathy for Eve's scepticism.[6]

Fortunately, scepticism and reason are also the key to equanimity – not just useful in life but also a source of happiness. I mentioned above how Thales' long stretches of observing the planets and contemplation yielded stunning scientific insights. I believe these long periods of thinking and reasoning must have been for Thales a source of joy and peace in themselves. I hope that by the end of this book, the reader will be persuaded that the pursuit of knowledge is a form of hedonism.

I want to emphasize that by stressing the value of reason, I do not wish to downplay actual mental illnesses where medical help is required. In those situations, one may have no ability to marshal one's mind and reasoning may not be of much help. Ironically, a good example of this is the American mathematician John Nash (1928–2015), among the most important personalities in game theory. Nash, one of the finest minds in history, was incapacitated for large parts of his life by schizophrenia, periods when this master of reason could not use his own faculties to help himself.

This book does not deal with these deep psychological problems. It is about negotiating the trials and tribulations of everyday life. Over the years, I have become convinced that reasoning is one of the most readily available and underutilized medicines.

Melancholia

I consider myself a person of generally happy disposition, with few anxieties, who enjoys company – of both men and women – and who is fortunate to have stumbled upon a career that he loves.

It was not always that way. I had a dark period, starting when I was seventeen years old, when despair gripped me and the world seemed hopeless. Harbouring all my anxieties inside me, I was sure this darkness would never lift. Luckily, it did. It began to improve, after about a year and a half, and eventually lifted totally. It was embarrassing for me to have such an unexpected episode and I did not feel like speaking about it. I did not tell my friends, and I did not tell my parents because it would have been too distressing for them.

I had led a sheltered, happy childhood in Kolkata, growing up in a household with parents and four elder sisters who doted on me. My father was an introvert and his love for his children had few behavioural manifestations. It was noticeable that every time my sisters or I were unwell, even with a common cold, my father spent longer than usual in the office. My mother laughed and told us he could not bear to see his children unwell, and so opted for a cowardly vanishing act (making frequent phone calls from his office to check how we were doing).

Growing up, I had a large clan of relatives spread through the city, who dropped in with a frequency that would shock a modern household. After I moved to the US, I remember one of my American colleagues, curious about Indian culture, asking me if my parents knocked on my bedroom door before coming in. Not only did my parents *not* knock, nor did our neighbours, Mr and Mrs Ghosh.

My father grew up in a poor household that plunged into even greater poverty after my grandfather died prematurely. My father struggled to earn a meagre income to help his mother and eleven siblings and got a law degree quite late in life by studying in the evenings. It was totally unexpected that he would do as well as he did as a lawyer. By the time I was old enough to have memories, it was a household with multiple staff, drivers and helping hands, loyal to us in a rather feudal way. In 1969, when I finished high school and was getting ready to continue with the sheltered life of living with my parents and being driven every day to college, my mother came to me one day and said my father wanted to know if I would agree to go and study in Delhi, at St. Stephen's College. It would just be three years away from home. It was obvious from the tone that my father was urging me to do so. This was a bolt from the blue. St. Stephen's College was India's premier undergraduate institution, but I had never thought my parents would want me to leave home until much later.

In retrospect, this was one of the wisest moves in my life. My father was a worldly philosopher. Though my departure would be painful to him, he had calculated that I needed the jolt of being on my own in order not to become a vegetative homebody. I agreed to apply to St. Stephen's.

There was a touch-and-go moment during the interview when I lied and said I wanted to do economics because I loved the subject, and the chair of the department, Mr N. C. Ray, asked me what I had read in the field of economics. I could not say I had developed the love from having read nothing. I fumbled, and then recalled that one of my school textbooks had a half-page on 'Marx's theory of surplus labour and why it is wrong'. I had studied at St. Xavier's, a Jesuit school in Kolkata, and our missionary moral-science teacher had made us learn that half-page by heart. I told the interview board I had been reading about Marx's theory of surplus labour and had reached the conclusion that it was wrong. I spoke for about five minutes elaborating on this. The interview board was impressed by my knowledge and I was impressed by my capacity to speak without knowledge.

And so, in July 1969, at the age of seventeen, full of trepidation, I left home for the first time to live in another city. I moved into my dorm at St. Stephen's College, in Rudra South. I met students much more boisterous than me, who had been educated in westernized boarding schools in different parts of India. The novel setting, being away from home and a feeling of inadequacy with these westernized classmates did create anxieties, but in ways that were normal, at least during the first three months.

After returning to college from the two-week October break, the plunge began. To this day, I do not know quite what happened to me. Was it triggered by moving out of a home where I felt totally, absurdly protected? Was it caused by a feeling of inferiority – a concern that I was not up to the mark with such smart classmates? Was it a specific psychological problem which had a name? Were there others who got it? Is it known what causes it?

For me, now, this is a matter of pure intellectual curiosity. In case one of my readers has an answer, let me fill you in with one or two details. One marked feature of this anxiety or depression or melancholia – I do not quite know what to call it – was its clear daily routine. I would wake up feeling more or less fine, then some time in the morning the anxiety would start, building up through the day, becoming acute by late afternoon. Then, as night descended, it would begin to ease.

As the months passed, the daily interval of calm, from night to morning, kept getting shorter. This growing anxiety and depression were accompanied by a loss of interest in everything.

It had no daily cycle; it was persistent. I had no ambition; I no longer cared for any of the things that had been dear to my heart. It was a cause of genuine despair that I appeared to be living with no purpose whatsoever.

I read that John Stuart Mill had a similar episode in his life when he was twenty. Normally, I would have been thrilled to find that I had something in common with John Stuart Mill, but at the time this too meant nothing. I continued to attend classes, had lots of friends, genuine and close. No one guessed what I was going through. After a year, I was fully reconciled to the fact that this pall of darkness would never lift from my life.

But it did. I do not know what got me out of it. I did see a psychiatrist in Kolkata in the middle of this, the only time in my life when I have done so. A well-read, cerebral person, he talked about Freud, Jung and others, and said that a lot of human problems arise from our ascribing too much importance to one aim in life – sex, money or fame. He said that, for people of my age, a lot of psychological stress arose from a latent sexual anxiety. He blamed Freud for this. Freud's emphasis on the sexual origins of our psychological problems became self-fulfilling. The psychiatrist argued that once we realize there is no single purpose or target in life that takes precedence over the others, it takes a huge load off our shoulders. I don't know whether his counselling helped me directly, but I remember his intelligent, humane conversation fondly.

The start of the lifting of the depression – I call it depression for want of a better word – from around the age of twenty was quite baffling, since by then I was reconciled to a life in its shadow. After another year or two it was gone. I did not speak about it for years. Partly out of shyness, partly a fear that, in talking about it, I might rekindle the dormant genie.

I do not know what made the melancholia go away and whether, like some episodic virus, it would have gone away anyway, no matter what I did. But there is one strategy which I began using around this time, which has stood me in good stead: reasoning with myself, and trying to be completely honest when doing so. Since I did not have access to antidepressants – almost no one in India at that time did – reasoning inside my head was my only ammunition. Whether or not it helped me specifically with my period of crisis, I emerged with the belief that honest, ruthless reasoning inside your head is one of the most powerful and underutilized recipes for happiness.

Take anger, for instance. There may be some value to *displaying* anger on occasion, but I am not talking about that.[7] I am talking about *being* angry. When we reason with ourselves, we will see that it is of no use. If when playing chess you get furious that your opponent moved the rook sideways, it is unlikely it will help you respond better. Grandmasters do not get angry when they are losing at chess; they focus their powers of reason. You do not have a choice over what others do. All you can do is to assess the situation and think of what *you* can do. That is what players do in game theory and that is what you should do in life. At times, a show of anger may influence how the other person behaves in the future; but your *feeling* of anger does not.

For some people, temperament alone can do what others need reason to accomplish. I am acutely aware of this because my father had a seemingly flawless ability for analytical reasoning, whether when arguing about the law or teaching me

Euclidean geometry, and that helped him greatly. My mother, however, was totally inept at deductive reasoning. Never lacking in confidence, she used to laugh and tell us how she failed all her school mathematics. What she had instead was an emotional robustness and an instinctive grasp of practical philosophy.

I have vivid memories of this from when I was quite young. A relative of mine, who lived and worked in Germany, died suddenly. The poor woman's mother, who was my mother's cousin and lived in Kolkata, was heartbroken and would come crying to our home. My mother, always kind, would stroke her arm and console her. This went on for a long time. Gradually, my mother grew puzzled about her cousin's persistent pain. One day, to my dismay, she asked her, 'I know this is a big tragedy, but your daughter was not living with you. She was in Germany. Can't you tell yourself she is still in Germany?'

The capacity to equate death with living in Germany showed a philosophical audacity that the Stoics would envy.

The Stoics, but not Archie, who in Tom Stoppard's play *Jumpers* memorably comforts Dottie, the wife of a professor of moral philosophy, who is distraught, having learned of an acquaintance's sudden death: 'There is no need to get it out of proportion. Death is always a great pity of course, but it's not as though the alternative were immortality.'[8]

Intelligence and emotions

The capacity to reason is the most valued and the most underutilized of human faculties. A lot of human woes, social and economic, stem from its absence. I will turn to some of these larger matters in later chapters. I want to begin with individuals and everyday life, and with our struggle to find happiness, success and acceptance in society.

Every time we feel unable to cope with the many demands of life, or a sense of failure about our careers or friendships, our instinct is to go to a psychiatrist, take medicine or, worse, shut ourselves in. I do not doubt that medicine is often needed, but I also believe that in many of these situations there is a substitute or at least a complement to medicine: reason. I am talking of simple deductive reasoning carried out mainly within one's own head. The art of reasoning well is critical for succeeding in the world of work, relating well to our social peers, fighting our eccentricities and even depressions. Unfortunately, we human beings are full of mental blocks. Especially under stress, we reason poorly and try to comfort ourselves in ways which are flawed and do not help in the long run.

To reason well in everyday life, you do not have to be a mathematician, game theorist or analytic philosopher or to have a high IQ. The ability to reason is a standard feature of *Homo sapiens* and is available to virtually all of us. The main stumbling block is that, when it comes to dealing with our personal problems, troublesome friends, nosey in-laws (the adjective may be redundant) or spiteful boss, or thinking about tense political and social matters, we reason less well than we are capable of.

I reached this conclusion after years of working with brilliant mathematical economists. It often took me by surprise how poorly so many of them reasoned as soon as they stepped outside their comfort zone of mathematical models and Greek symbols, to talk about personal matters, office politics or world affairs. I do not want to lose friends, so the names of these economists will have to remain undisclosed.

These interactions led me to conclude that the failure to reason in real life is, in large measure, an emotional failure. To test this hypothesis, I conducted a series of informal experiments on my students at the Delhi School of Economics and, on some occasions, on ordinary people in downtown Ithaca, with the help of my game-theory students at Cornell. Here is a sample of the kind of experiment we did. This was not scientifically controlled and so I did not try to publish it (I hope some of the readers of this book will be tempted to do such tests more formally, with proper statistical controls).

Here is the basic idea. Ask someone to look at three statements, A, B and C, assume A and B are true, and then, based on that assumption, say whether C is true or false. Do the first experiment with the following bland statements:

A: All men wear hats.
B: Tom and John are men.
C: Tom and John wear hats.

Almost all respondents get this right. They say 'true'. This kind of logic is known as a syllogism and it has come down to us from the time of Aristotle. There is also clear reference to this kind of logic in the early Vedic writings on *Nyāya*, a school of thought that goes back to the second century BCE in India. You begin by asserting that a set of creatures has a trait. Then you note that a creature belongs to the set. Almost all human beings can deduce from these statements that the creature has the trait.

Interestingly, most people intuitively understand syllogisms and get the right answer, *as long as the words and topics are not emotive.* To test this, we can create another experiment, which is also a syllogism and has an identical logical structure to the above test involving hats, but this time it deals with less innocuous matters. Let me call the statements A', B' and C' to distinguish them from A, B and C.

A': All creatures are fit to do only housework.
B': All women are creatures.
C': All women are fit to do only housework.

As before, we asked respondents to assume that *A' and B' were both true*, and say if C' is true or false. What we found was that, despite being asked only to assess the logic and not the value of the statements, a much larger number of people got the deduction wrong. Many more would say C' is false than would say C is false in the previous experiment.[9] Poor reasoning is often a sign not of low intelligence but of psychological barriers. The human ability to reason falters when humans have to apply logic to emotive matters.

I should point out here that 'if–then' statements of the above kind are what we call 'positive' statements. There is no moral or normative content in them, no implication about how we believe the world should be, no recommendation of how to behave. In the above example, your asserting that *if* A' and B' are true, *then* C' is true, says nothing about whether you are a sexist or feminist, whether you are against or in favour of big government or whether or not you believe that there should be higher taxes for the rich and food subsidies for the poor.

We might challenge one of the premises – are all creatures really fit to do only housework? But that does nothing to affect the logic of the propositions.

Other examples of positive statements are propositions like '2 plus 2 is 4' or '7 plus 7 is 15'. They can be true or false, but they cannot be moral or immoral. David Hume, one of my two biggest influences as a philosopher, was basically pointing this out when he argued that you cannot derive any normative conclusion from a set of purely positive statements. This is referred to as Hume's law.[10]

Hume's law is not just important for philosophers and philosophical debates, but critical for good decision-making.[11] We often muddy our thinking on purely positive matters. Did slavery help boost US GDP growth? The answer to this question does not compel us to take any particular normative position. It is perfectly reasonable to say that, no matter what the answer to the question is, slavery is a shameful episode of history and should not be allowed ever again. Since our analysis on purely positive matters has no consequence by itself on what kinds of actions we choose, it is important to do our thinking on positive matters with regard to nothing but the truth.

This is what Bertrand Russell (my other favourite philosopher) famously said when, in a 1959 BBC interview, he was asked for his advice to future generations:

> I should like to say two things, one intellectual and one moral. The intellectual thing I should want to say to them is this: When you are studying any matter or considering any philosophy, ask yourself only what are

> the facts and what is the truth that the facts bear out. Never let yourself be diverted either by what you wish to believe, or by what you think would have beneficent social effects if it were believed. But look only, and solely, at what are the facts.
>
> The moral thing I should wish to say to them is very simple. I should say love is wise, hatred is foolish. In this world, which is getting more and more interconnected, we have to learn to tolerate each other . . .

Essentially, he was appealing to Hume's law. In our analysis we should make no compromises. By the same premise, in deciding what is morally right, we need not be constrained by the if–then propositions that we may have derived.

One of the most moving illustrations of Russell's advice of total commitment to the truth being taken seriously was his daughter, Katharine Tait. Katharine grew up to reject many of her father's moral stances. Bertrand Russell was an atheist (most of the time; occasionally he called himself an agnostic), who rejected Christianity. Katharine became a devout Christian. Her biography of Bertrand Russell catalogued a lot of the pain she suffered, growing up in an intellectually avant-garde home, and her father's multiple marriages and love affairs. The biography ends with these poignant lines. I am quoting the last paragraph in full:

> It was a long time before I thought of writing about him. I will tell the world what a great father he was, I said to myself, how wise and witty and kind, how much fun we always had. They mustn't think he was a cold rational

> philosopher. So I thought and so I began to write, but it has not come out that way. The 'but's and the complaints seized my pen and forced it to record them. 'He loved truth you know,' they urged, 'You cannot honour him with a lying memoir. You must set down all that was wrong, all that was difficult and disappointing, and then you can say: He was the most fascinating man I have ever known, the only man I ever loved, the greatest man I shall ever meet, the wittiest, the gayest, the most charming. It was a privilege to know him and I thank God he was my father.'[12]

I must clarify one topic on which Russell was not always clear. There is a distinction between truth and *telling* the truth. We should try never to compromise on seeking the truth, but we may occasionally have to compromise on telling the truth, for moral reasons. Saying something is an act. It is an act of speech and, like all acts, it can wound and hurt, and it can be morally right or morally wrong.

Consider a kind old man who invites you to dinner, spends a large amount of time cooking, and at the end of the meal asks if you enjoyed the food. Enjoying or not enjoying is a fact that can be true or false. But *saying* you enjoyed or you did not enjoy is an act. It is entirely in keeping with Hume's law to decide in your head that the food was awful but to say, 'It was delicious; I loved it.'[13] Indeed, this may be the right thing to do, for the *moral* reason of not hurting someone's feelings needlessly.

More generally, since public statements can hurt and wound, as well as trigger unseemly actions on the part of others, there may indeed be a moral case for not stating some truths aloud. Within your own head you have to allow the free examination of all propositions and be totally honest about your findings. This is critical for good decision-making and even for happiness, because so much of human unhappiness stems from our own confusion.

Allowing extraneous considerations to cloud our judgement concerning the validity of propositions leads not just to bad individual decisions but is also responsible for many big policy mistakes made by nations. Doomed are societies in which people, after hearing about Pythagoras' theorem, want to know about Pythagoras' political party affiliation in order to decide if the theorem is right.

Bertrand Russell, in his *History of Western Philosophy*, defended Machiavelli by pointing out that Machiavelli's 'recommendations' to the prince or the political leader – which people take to be immoral – were in fact not immoral because they were, according to Russell, not really recommendations but if–then statements. Machiavelli was essentially saying, '*If* you want to survive as a leader, *then* you should do this.' This is different from saying, 'You *should* do this.' The latter is a normative statement and can be moral or immoral. The former is a positive statement. According to Russell, Machiavelli was merely trying to unearth the laws of politics and not recommending any particular behaviour.

Despite this, when the words we use touch emotional chords in our heads, many of us have difficulty in thinking clearly. That lack of clarity can allow others to exploit us. This comes out clearly in Voltaire's alleged quip, 'There is no God, but don't tell that to my servant.'

Voltaire may have said this in jest, but the method of deluding others about God predates him by quite a margin. Kautilya, the Machiavellian adviser to the founding emperor of the Maurya dynasty, which ruled India in the third century BCE, wrote about a variety of such methods. In his classic book, *Arthashastra*, advising the king about economics and political control, he points out how religion and the fear of God can be used by the king to make ordinary citizens do what the king desires.

Kautilya proposes that if the king's coffers are empty and the state's budget is in trouble, the king could resort to organizing fake 'miracles', like building a temple quickly in a remote area and telling people it appeared spontaneously, and then encouraging people who will be dazzled by such events to give offerings of money to God. The money thus collected can be used by the ruler to close the nation's fiscal deficit. (Of course, once the king masters this art of deception, he can use it for other, less stately, purposes as well.)

The world is your laboratory

The power of reasoning, which often sits dormant in us, is closely intertwined with game theory. As we have seen, this beautiful discipline that is so important in so many fields is surprisingly young. There were a few scattered theorems proved by mathematicians starting barely a hundred years ago. There was important path-breaking research done after the 1940s, including John Nash's seminal papers in the early fifties. Then, in the early sixties, came Thomas Schelling's classic book, *The Strategy of Conflict*. This and his later work, using no mathematics but pure deductive reasoning, illustrated how adaptable game theory is. It has room to house pure mathematicians as well as thinkers without a shred of mathematics, both of whom can make deep contributions to different aspects of the discipline.

With game theory, you also do not have to go anywhere to be in the laboratory, because we live in the laboratory. All we need to do is be curious: to scrutinize the world and pick up cues from everyday life, and string them together with logic and reason to discover new patterns and new truths. Game theory is about rational behaviour in situations where others, including your enemies, have their own desires and motivations. We are all trying to further our own interests at the same time. In the words of one of the prominent figures of the discipline, Robert Aumann, it is all about 'interactive rationality'.

Game theory is morally neutral. It does not tell us what individuals should aspire to achieve but analyses how they can get what they want, and describes the consequences. It can be the story of some corporations competing for customers, each with the aim of maximizing its own profit, but it can also be a tale of Good Samaritans trying to create a better world. As a result, it helps you reason more clearly about how you should go about achieving your objective, *whatever that may be.*

But I want to remind the reader of the importance of morality. Morality got banished to the sidelines in mainstream economics because of a surprising discovery by Adam Smith, and its widespread misunderstanding and misuse.

Smith showed that, in many situations, you do not need to *be* good to *do* good. A group of individuals, motivated by nothing but their self-interest, can lead society, almost as if guided by an 'invisible hand', to an outcome that is collectively optimal. The publication of this idea in Smith's most famous book, *The Wealth of Nations* (1776), would have a huge influence on the profession of economics and would eventually spill over into the thinking of everyone else. So powerful was this idea that 1776 has gone down as that rare year which is celebrated equally for two reasons – the declaration of independence of a nation, namely, the United States, and the publication of a world-changing book.[14]

As if the coincidence of that year were not enough, David Hume chose 4 July 1776 for a party in his home for his friends, including the most prized, Adam Smith. The reason the party was so special is that he organized it to celebrate the end of his life, since he knew he was dying of cancer. Hume died less than two months later, on 25 August.

Hume has lessons for us not just through his intellectual contributions but, like some of the early philosophers, such as Epicurus, and even political leaders, such as Marcus Aurelius, through his way of life. People of such immense talent are often not pleasant personalities. Hume was an exception. A radical in his day, often attacked for rejecting conventional religion and wisdom, he was, nevertheless, a person of remarkable equanimity. Adam Smith wrote in a letter to William Strahan, Hume's publisher, dated 9 November 1776, that as Hume's health had deteriorated, he had told Smith, 'I am dying as fast as my enemies, if I have any, could wish, and as easily and cheerfully as my friends could desire.' (Rasmussen, 2017.)

Smith's idea of the invisible hand was deep enough that its full formalization and understanding would take much more time. Could Smith's idea of the invisible hand be 'proved' formally, the way Euclid had done with many of our intuitive ideas about shapes and geometry? The answer turned out to be yes, but it took nearly two centuries to get there. The full description of the economy, with supply and demand for every good specified, emerged in the late nineteenth century, with the pioneering works of William Stanley Jevons, Léon Walras and several others. Showing under what circumstances this large system would have an equilibrium, which would then exhibit the property of the invisible hand, would take us into the mid-twentieth century and involve the seminal work of Kenneth Arrow and Gérard Debreu (1954).

The excitement of this new idea of an invisible hand made us forget that, though the connection between individual selfishness and the collective good was important because it was intellectually so surprising, it is not the only prerequisite for achieving the collective good. In particular, these include individual morality, such as the instinct not to betray other people's trust and to have concern for others. This is the morality of the Enlightenment thinkers. I am not talking about making sacrifices and being good in order to increase the chances of going to heaven, as some religious instructors advise. That can play a useful functional role, but it is not morality; it is cost–benefit analysis.

I point this out while being aware that there are many practitioners of religion who try to do good for its own sake and share the values I am upholding here. I do not believe in God. The morality that is important to me is the one that stems from other-regarding behaviour; it includes respect for individual freedom – giving people the freedom to do what they want as long as it does not hurt others. The morality that is important to me is the nurturing of basic kindness and compassion, of treating all human beings equitably, irrespective of race, religion, nationality, sexual orientation and other such markers of identity. We do not have to go far back into history to discover our common origins and realize how superficial some of these differences are.

If the drive for individual betterment is the fuel that drives the economy, our morals are the nuts and bolts that hold this complex machinery called the economy in place. In a short 1983 paper with a long-winded title, 'On Why We Do Not Try to Walk Off Without Paying After a Taxi Ride', I argued that we do not do so because, beyond our urge to do better for ourselves, we also have some hard-wired codes of conduct. The modern economy would not survive if we had to rely on the police and the courts for every contract to be enforced, such as the implicit contract that, after the taxi takes me to my destination, I must not run away without paying.

It is easy to run away without paying after a taxi ride. If everybody did so, however, there would be no taxis plying our streets. It is because we have the basic moral value of doing our share after the other person has done his or her share that we – most of us, that is – pay up. It is that assurance that persuades the taxi driver to give us the ride in the first place.

A surprising number of mainstream economists do not like to admit that anybody does anything because of morals. They argue that the reason we pay is not morals but because taxi drivers are often big and burly and can beat us up if we don't pay. What these neoclassical economists fail to see is that, if their argument were right, taxi drivers could easily threaten you to get a second payment. By this argument they would manage to collect (at least) twice the meter reading. The fact that taxi drivers (barring maybe a few neoclassically inclined ones) do not typically behave in this way shows that people do generally adhere to norms. If they did not, no passenger would take taxis and the taxi system would break down.[15]

Indeed, there are differences across societies in terms of how strongly people adhere to these norms. As a result, the taxi market functions more efficiently in some societies than others, thereby showing that economic markets depend on more than just individually rational behaviour.

The very fact that so many markets function reasonably smoothly indicates that we do adhere to a modicum of moral and social norms,[16] even though each of us could do better by unilaterally violating them.[17] It's a reminder of the need to live by these norms and nurture your natural altruism. Some of these values will conflict with one another, causing moral dilemmas. I shall address some of this later. While this is basically an amoral book, showing you how best to achieve whatever you wish to achieve, I hope that you will wish to achieve not just what is enriching for yourself but what is good for society as a whole (and also for creatures that cannot speak). The best way to reinforce this is to quote Bertrand Russell's morally resonant lines, published in the *Nation* on 15 August 1914, on the eve of the First World War. Reminding his fellow citizens of their larger commitment to humanity, he wrote:

> The enemy are men like ourselves, neither better nor worse – men who love their homes and the sunshine, and all the simple pleasures of common lives. And all

> this madness, all this rage, all this flaming death of our civilization and its hopes, has been brought about because a set of official gentlemen, living luxurious lives, have chosen that it should occur, rather than any of them should suffer some infinitesimal rebuff to his country's pride . . .

War, of course, is the ultimate game. The reasoning in this book teaches you how to win. However, to know how to win does not mean you have to win. There are many games in life where it is morally incumbent on us not to try to win.

In the last three chapters, we will explore matters of collective human concern, such as climate change and war. How can we try to avert these dangers? This goes beyond self-help, to group-help. Interestingly, and paradoxically, this urges us to use game theory to go beyond game theory. It makes us ask not just how we may play the game of life, but how we might change the rules of the game.

2

Game Theory in Everyday Life

Is jogging worth it?

Game theory is the analysis of interactive rational behaviour: being rational, while taking into account other people's rationality. Since the first requirement is such an integral part of traditional social sciences, it is the interactive part that is special to game theory. The rationality of others is more complex than appears at first sight because how other people choose will in turn depend on their belief about how you will choose.[1]

However, even demanding our own rationality is not trivial. It requires that, when making choices, you keep in mind all the actions or strategies open to you, and then choose the action that maximizes whatever it is that you are striving for (the 'payoff'). Game theory has no opinion on what gives people their payoffs. That is entirely up to them. Once it is determined, players choose from among the available actions (the 'feasible set') to maximize their payoff. There are times, however, when we get carried away by an immediate impulse or habit and get distracted from our larger purpose.

I got a reminder of this many years ago, in a relatively trivial way. I gleaned somehow – I forget the original source – that every ten minutes of jogging that you do increases your life expectancy by eight minutes. That sounded like a wonderful deal. Jog for half an hour on five days one week, and your life expectancy will increase by two hours; that is, you can expect to live for two additional hours on Earth. Now keep this up every week and your life expectancy will go up in leaps and bounds. It sounded such a good deal that I began jogging fairly regularly.

Then one day, while out jogging, I asked myself what my aim in life was. Or, in the language of game theory, what was my payoff function? Did I want to maximize my time on Earth, or maximize my *non-jogging* time on Earth? If it was the latter, then every time I jogged for ten minutes, I was losing two minutes of my non-jogging time on Earth. I almost stopped in my tracks. Since then, I have started and stopped jogging several times, not because of irrationality but because I have vacillated over what my true payoff function is.

The message is clear. You need to know what your payoff function is before making the right choice. This is a simple axiom but worth reminding ourselves, especially when we confront difficult decisions in life where it is easy to lose sight of the larger picture. It is our occasional propensity to lapse on this that opens up the room for counselling and self-help manuals.

For decision-making amid other players (or 'agents') pursuing their own payoffs, there are further complications.

Situations arise every day whereby you have to take account of the fact that others also are thinking creatures with their own aims and ambitions. This is true in war and diplomacy, as we have already seen, as well as in competition between firms and corporations. It is also true outside Indian temples.

Visiting temples, as a child, was a somewhat harrowing experience because there were often thieves lurking amid the milling crowds. When entering an Indian temple, you are expected to remove your shoes, which, in those days, would usually be left in a heap outside. It was a common experience that after your visit you would come out to discover that (while you may have been blessed by God in other ways) your footwear was gone.

In high school I devised an effective technique to foil the thieves. Indian temples usually have multiple doors. The method I used was to put down one shoe in the pile of footwear at one door, hobble over to another door and set down the other shoe there. Then I could go in and pray or admire the statuary to my heart's content without the nagging worry of whether my shoes would be stolen and I would have to go home barefoot. I taught my siblings and cousins about this method and it was hailed as a secret to footwear riches in our larger clan.

I had thought about what I wanted, then thought one step further to what the thieves wanted. If, however, the thieves had thought about what I might be doing to deter them, the efficacy of my strategy would surely have gone down, because thieves would have learned to pick up nice shoes, even if they were not in pairs, and rush to other doors to complete the set. Luckily, we managed to keep this technique a closely guarded secret.

One of the most important lessons from game theory is not to underestimate the rationality of others. Consider the tale of a hat-seller, with his big collection of hats, walking in some remote part of India from one village to another, when he felt sleepy.[2] So, he decided to set down his wares under a shady tree and take a nap. When he woke up, to his dismay, all the hats were gone. He discovered that monkeys had come down from the tree to investigate, had taken up the hats and were now wearing them. He was a poor man and in his desperation and frustration, he took off his own hat and threw it down on the ground. Now, monkeys, as we know, are great imitators. Soon all of them were throwing down the hats. The hat-seller was relieved. He collected the hats and went on his way.

Forty years later, his grandson, who, in keeping with the family tradition, had become a hat-seller, was going with his wares from one village to another and felt like a nap. So, he set the hats down and went off to sleep. When he woke up, he discovered that monkeys had taken them to the treetop and were wearing them. He was not a rich man; what could he do? And then he remembered his grandfather's story. Relieved, he took off his hat and threw it down. One solitary monkey clambered down, picked up the hat, put it firmly under its arm, walked up to the hat-seller, gave him a slap and said, 'Do you think only you have a grandfather?'

The lesson from this tale is the essence of game theory. We shall encounter examples in this book where being rational but failing to take account of the rationality of all other agents results in flawed decision-making. A lot of government welfare programmes aimed at civilian behaviour go wrong because governments design them without taking account of the fact that bureaucrats – civil servants, the police, the magistrate – meant to carry out the enforcement are also rational agents with their own aims and motivations. This is a flaw that underlies a lot of traditional economics. The so-called Chicago school of law and economics assumed that the law affects behaviour because it changes the payoffs we receive from certain actions. When there is a new speed-limit law, a person considering driving fast will, as before, consider the risk of an accident, but now, in addition to that, they will calculate the probability of being stopped by the police and fined for speeding. It is this additional cost that changes behaviour. The police officer is assumed to be a robot or a saint who mechanically does what the law requires.[3] This flaw means that mainstream economics is deficient in explaining why the law is often not enforced and why, in many countries, after a person is caught breaking the law, they get away with not paying the fine but paying a smaller bribe instead, which goes into the pocket of the police instead of the state coffers.

We owe a lot of our failures in life to our inadequacy in putting ourselves in the shoes of others. One of the main lessons of this book is to conduct this exercise in empathy as accurately and dispassionately as we can. It may bring surprising benefits.

When flying, most of us would welcome having an empty seat next to us. You can set down a book there, or your pillow or laptop, or simply enjoy the extra elbow room. When you travel with a partner or a friend, you want to sit close, but having an empty space between you is an additional bonus.

So here is how you book your seats. Look for three empty seats in a row. If a row is available, book the two outside seats. As the plane gets full, there is a chance that the one seat between you and your friend will remain vacant. If that happens, you can use that as the vacant seat next to each of you, or one of you can move over to enjoy each other's company and the one free adjacent seat.

There is, of course, the possibility that the plane will be fully booked and someone will eventually get the seat between the two of you. If that were to happen, you could ask that person to change with one of you. Most people hate sitting ensconced between two talkative friends, meaning that person is motivated to agree to the switch. So you either gain or draw.

There is a small possibility that the person between the two of you turns out to be someone who actually likes to sit between others. In that case you lose, but this has such a low probability that it is safe to ignore it. And if this happens, treat it as a reminder of the philosophy of scepticism – that there is nothing certain in life.

The origins of game theory

In 1838, the French mathematician Antoine Augustin Cournot published his now-famous book, *Recherches sur les principes mathématiques de la théorie des richesses*, which developed ideas about equilibrium as part of an effort to understand the nature of competition among a limited number of producers of some good. Some of these ideas were later refined in a book review of Cournot's work written by another French mathematician, Joseph Bertrand, in 1883.

The basic idea underlying Cournot's path-breaking work is simple. In deciding on a strategy, each player will keep one eye on what strategies the others are using. If, knowing how others will choose, each player is happy to stay with their own strategy, we reach equilibrium. No one will individually want to deviate from this.

Bertrand's book review was historic for two reasons. First, the time lag between the publication of Cournot's book (1838) and the publication of the book review (1883) – a full forty-five years – must be something of a record. Second, I am not sure there is any other book review that gave birth to a concept named after the reviewer, in this case, the 'Bertrand equilibrium' which shifted attention from quantity to price.

Ideas about equilibrium turn up in early works on board games, like chess and bridge. One of the most fascinating theorems was proved in the context of chess by the German mathematician Ernst Zermelo, in 1912. The basic idea is simple enough. In chess, sometimes the first-mover (with the white pieces) wins; sometimes the second-mover (with the black pieces) wins; and on occasion there is a draw. What Zermelo proved was that this variety of outcomes happens *only* because our capacity to think is limited. Indeed, even the most sophisticated computers that play chess have limited capacity to analyse and compute. If, however, two *perfect* players played chess, Zermelo's theorem assures us, the outcome would be either white would *always* win, black would *always* win, or there would *always* be a draw.

It is important to understand that the theorem does not say that, in every game, one of the three outcomes will occur. That is completely obvious and trivial. You do not need a mathematician for that. What Zermelo proved was that, with perfect players, once you made them play and saw the outcome, you would know exactly what would happen in all future games. Chess is an exciting game only because our capacity to reason is limited.

In the twentieth century, there were major mathematical breakthroughs culminating in the publication of John von Neumann and Oskar Morgenstern's book, *Theory of Games and Economic Behavior*, in 1944. This book outlined a full theory of games and attempted to connect the mathematical structure of game theory to the social sciences.

After this, in 1950 and 1951, came three papers by John Nash which quickly became seminal, establishing theorems which were critical in the founding of both cooperative and non-cooperative game theory. John Nash is, arguably, the most important figure for modern game theory.

My involvement in game theory has always been tangential. I worked in other areas and picked up game theory in bits and pieces, with special interest in its philosophical moorings and the paradoxes that it threw up. The discipline was still barely mentioned in the early 1970s when I was a student at the London School of Economics. My doctoral adviser, Amartya Sen, who would later win the Nobel Prize in Economics, was an exception; and I did learn some elements of the theory from his lectures.[4] But when it comes to game theory, it is John Nash who has a special place in my heart.

On 23 May 2015, as my wife and I were driving back to Washington DC after a weekend in Virginia visiting James Madison's home and the birthplace of the American constitution, our daughter called to give us the news. John Nash and his wife, Alicia, had just been killed in a car accident on New Jersey Turnpike. The brutality of the news was difficult to fathom. How could a person of such genius, after a life of so much struggle battling schizophrenia, go in such a banal way?

The news of Nash's death brought back memories of our first meeting. In 1989 I was a visiting professor at Princeton, on leave from the Delhi School of Economics. By then, Nash's schizophrenia was in remission. He could be seen strolling the Princeton lawns for hours on end. For the residents of Princeton he was part of the scenery, and so unremarkable. For me and my colleague Jörgen Weibull, also a visiting professor, it felt strange. There we were in classrooms, analysing or applying the 'Nash equilibrium' and the 'Nash bargaining solution', and the man after whom these concepts were named was right outside, pacing the yard.

It was Jörgen who managed to get Nash to join us for lunch one day in one of the Princeton cafeterias. It was exciting to be in the company of a genius, even though he spoke little and, every now and then, seemed to drift off into his own thoughts. One vivid memory sticks in my head from that lunch. It was the reaction of a friend and economist, who would later win the Nobel Prize in Economics, Abhijit Banerjee. Seeing Jörgen and me there, Abhijit came and joined us. As we introduced him to Nash, it was like watching a young literature student who is about to sit down with friends for lunch and been told that the third person at the table is William Shakespeare.

Nash's most important papers were written before he was twenty-five, and his creative period was over by the time he was felled by schizophrenia in his late twenties. By the age of thirty-two he would spend long stretches of time in psychiatric hospitals, and the next thirty years were a battle with paranoia and delusions. It may be pointed out here that, for Nash, the remission from schizophrenia was not unmitigated good news. As he observed in the short autobiographical essay he wrote after he won the Nobel Prize in Economics in 1994, the return to rationality 'is not entirely a matter of joy as if someone returned from physical disability to good physical health. One aspect of this is that rationality of thought imposes a limit on a person's concept of his relation to the cosmos.' He went on to talk of how limited the prophet Zarathustra would have been without what others referred to as his 'madness'.[5]

John Forbes Nash, Jr., was born on 13 June 1928, in Bluefield, West Virginia. He was recognized early as a prodigy; he got his PhD in mathematics from Princeton at the age of twenty-two. His PhD thesis was as short as his productive life: a mere twenty-eight pages. One of his most celebrated papers, which was part of his PhD, lays out the conditions for the existence of a 'non-cooperative equilibrium' in games, in just one page and a few lines. It's so admirably concise that the entire paper was reprinted on a T-shirt designed by graduate students in the economics department at Cornell in the late 1990s. (I have a confession to make here: the first time I read the paper was on the T-shirt.)

Of Nash's many contributions, the one that got maximum play is the concept of the 'Nash equilibrium', which is used to understand the behaviour of oligopolistic firms, movements in financial markets, political rivalry and strategizing in conflict, such as during the Cuban Missile Crisis. We will learn more about the Nash equilibrium later.

The second time I met Nash was after he had become a worldwide celebrity. He had won the Nobel Prize, sharing it with John Harsanyi and Reinhard Selten, in 1994, and was the subject of a popular Hollywood film, Ron Howard's *A Beautiful Mind*, in which he was played by Russell Crowe. He was part of a conference in Mumbai in January 2003 that brought several prominent economic theorists to town, including Robert Aumann, Roger Myerson and Amartya Sen. The audience for Nash was large, with some recognizable faces from Bollywood, who may have come expecting to see Russell Crowe. The talk was disappointing, as Nash tried to address some practical policy questions. He was too much of what the philosopher Isaiah Berlin calls a 'hedgehog': he was good at focusing on one thing very deeply but, unlike the fox, he wasn't able to range over many topics, as he tried to do in the talk.

The next day, much to my surprise, as I prepared to give my own talk, Nash came in and sat in the front row. I was thrilled but nervous. How would I talk about the Nash equilibrium with Nash himself sitting right in front of me? I need not have worried. He fell asleep within five minutes and did not wake up till the end of my lecture. After I finished, he strode up to me. We have all heard stories of those geniuses who can follow you through their sleep and ask penetrating questions. I readied myself for what he might ask and was excited about the deep insight he would offer.

He asked me, 'Can you tell me where the men's room is?'

That was the last I heard from John Nash.

Prisoner's Dilemma

In the early 1950s Albert Tucker, mathematician at Princeton, was visiting Stanford's mathematics department. There had been a mix-up in terms of space allocation and he ended up using an office in the psychology department. There he would be, door wide open, hunched over paper, scribbling and scrawling tirelessly. One day, one of the Stanford psychology professors dropped in and said that he and his colleagues were curious about what it was that Tucker did and asked him if he would care to give a seminar.

Tucker readily agreed. He was working on a game that had been invented by Merrill Flood and Melvin Dresher at the RAND Corporation, the non-profit organization set up after the Second World War for research especially in the context of military strategy. A game is defined in terms of three components. First, there must be a defined set of players. Second, each player must have a defined set of strategies or feasible actions to choose from. Third, each player must have a well-defined payoff function, which specifies the payoff or utility that the player gets after everybody has chosen their strategy or action. But presenting it in terms of pure numbers or symbols would not hold the attention of the psychology department. He realized he needed to lay it out as a story, and so he came up with one he called the Prisoner's Dilemma.

Tucker's Prisoner's Dilemma tale goes as follows. Two people have been arrested for an alleged joint crime and are being kept in separate cells. A magistrate trying to decide on their punishment hits upon an ingenious idea. He tells both prisoners that they each have to write down on a piece of paper 'confess' or 'don't confess'.

If both confess, the story is out and they will both be sent to prison for ten years each.

If neither confesses, there is still some independent evidence that they have done it and so the magistrate will send them to jail for two years each.

If one of them confesses and the other does not, then the one who has not confessed has not only committed a crime but is also now committing perjury. So the magistrate will send her to jail for twenty years. The other prisoner may have committed a crime, but is being so cooperative that the magistrate decides to let him off free, with no jail time. (We are not here to analyse the magistrate's sanity.)

The dilemma is: how will the prisoners behave when faced with these options? How each prisoner fares depends not only on what each prisoner chooses but also on what the other prisoner chooses. What happens in the Prisoner's Dilemma comes as a surprise, at least at first sight, and this is why this game is so celebrated.

To see what happens, put yourself in the shoes of one of the prisoners. Assume first that the other prisoner is going to confess. What should you do? Clearly, you should confess, because that way you will get ten years in jail, whereas if you do not confess, you will get twenty years. Next, suppose the other prisoner is going to opt not to confess. Then, surely, you should confess. That way you will get zero years in jail, whereas if you opt not to confess, you will get two years in jail. In short, no matter what the other player does, you are better off confessing to the crime. With both players reasoning this way, they will both opt to confess and each spend ten years in jail. There is nothing either player can do unilaterally to change this. In short, both confessing to the crime is the only equilibrium.

The Prisoner's Dilemma is a tragedy. If both players chose

not to confess, they would both be better off, with only two years each in jail. By being individually rational and self-interested, they ironically end up hurting their self-interest. This is why the game acquired such iconic status: it is a stark example of how pure self-interest exercised by all individuals may not in the end serve the self-interest of anybody. It finally gave us a word of caution about Adam Smith's invisible hand. While the invisible hand does indeed work in some circumstances by leading selfish individuals, unwittingly, to an optimal outcome, it is folly to think that that will always be the case.

Life is full of contexts where the Prisoner's Dilemma is relevant. Climate change and environmental damage figure prominently among them. If each of us guards our own self-interest and takes decisions on that basis, we are likely to continue to do enormous damage to the environment. Greta Thunberg is right when she expresses dismay about such selfish behaviour. When someone burns coal, a negligible part of the cost is borne by the one burning the coal, because the smoke goes out into the atmosphere and affects everybody.

In the popular media, we often hear people being told that taking individual action to stop environmental damage is in their own interest. This sounds good, but, sadly, it is not true. Similarly, taking action to prevent environmental damage will typically not be in your own interest, because you are too small a player. This is what is sometimes called a tragedy of the commons: we might all lose a public good because, as individuals, it is rational to take what we can get. The need here is to look beyond one's immediate self-interest.

Another striking example of the Prisoner's Dilemma is the

arms race. For each nation, individually, it is typically worthwhile to build up its arsenal of destructive weapons, with the aim of outgunning potential competitors. But when all nations do that, they are back to square one (with possibly the additional risk of a nuclear war), with a huge amount of resources diverted from necessities to making and storing weaponry. We can only break out of this by introducing new rules to the game in the form of pacts and contracts across nations.

The Stag Hunt

Another game that is worth having in our artillery of analysis is the Stag Hunt, also known as the Coordination Game or the Assurance Game. The Stag Hunt goes back to the mid-eighteenth century, to the philosopher Jean-Jacques Rousseau's *Discourse on Inequality* (1755). There are many variants of the game. Let me present here the central idea.

Consider a game with ten players. They are going on a stag hunt. Each of them has to choose between hunting the stag (S) and going for a hare instead (H). Each player has to decide on his or her strategy as an individual. Hunting the stag is not easy. Only if all ten players opt to go for the stag hunt will they succeed. So here are the payoffs. If all ten choose to hunt the stag, they take it down together, and each player gets $8 in reward. If all choose the solitary activity of hare hunting, they get much less, namely $2, but there is at least no chaos. If some choose S and some H, there is chaos in the woods with a stag running haywire, and no one gets anything.

It is intuitively easy to see that there are certain equilibrium outcomes of this game. Once we reach one of these outcomes, no individual can get more out of the game through unilateral action. One obvious outcome is where every player chooses S. They will all get $8 each and no one will have an interest in deviating, because by doing so the person will get zero. Another such outcome is where everybody chooses H. They will all get $2 each. Each player may lament that the group is stuck in a bad equilibrium, but there is nothing anyone can do as an individual to break out of this. Any individual deviating from H will end up getting zero.

There is one last class of outcomes where no one can do better through a unilateral change. Suppose two or more players choose S and two or more players choose H. They will all get nothing, but there is nothing anyone can do individually to lead to a better result. This is the outcome of anarchy, but an anarchy where each individual is helpless. This kind of game has been of interest to economists and philosophers trying to grapple with problems of collective action and leadership that go back to Hobbes, Rousseau and Hume.[6]

The Assurance Game or Stag Hunt has one advantage over the Prisoner's Dilemma. Once you have all players playing S, and have reached a position of 'good equilibrium', there is no need for any outside force to maintain the status quo. It is in each individual's interest to remain there. In a simplified two-player version, clearly there are two equilibria, (S, S) and (H, H). In the Prisoner's Dilemma, both players can do well by cooperating, but this outcome cannot be sustained without external force, since each player could do better by deviating.

Up to now, we have been talking about games where players move simultaneously, or, more generally, each player makes a move without knowing what the other player has chosen. We can, however, create variants of such a game by bringing time explicitly into the picture and having situations where one player moves first and then, seeing this, the other player chooses their action. Games in which players make moves in a sequence are referred to as extensive-form games. And games where players move simultaneously, such as the Prisoner's Dilemma and the Assurance Game, are referred to as strategic-form games or normal-form games. In what follows I shall rarely use these technical terms, since it will be obvious from the context what we are talking about.

We can think of a variant of the Stag Hunt where this will be explicit. Consider the same game as above but with only two players. Suppose the payoffs are the same, but Player 1 makes the first choice between S and H – and then, after seeing this, Player 2 will make his choice. It seems pretty easy to guess what will happen in such a game. Player 1 can easily see that if she chooses S, Player 2 will also choose S; and if Player 1 chooses H, Player 2 will also choose H. Knowing this, it is obvious that Player 1 will choose S.

Since we are already treading into the concept of equilibrium, it may be useful to introduce the idea more formally.

Nash equilibrium

The idea of equilibrium owes much to the seminal work of John Nash that gave it a formal definition and name. The definition of Nash equilibrium is simple. Given any game, after all players have chosen their actions, if each player finds that he or she cannot do better by a unilateral deviation to some other action, then this set of actions chosen by all the players constitutes a 'Nash equilibrium'.

In a game like the Stag Hunt, this means that after all players choose between S and H, if it turns out that no single player can do better by changing her action, *given the choices of all other players*, then we have reached a Nash equilibrium.[7] In the two-player version, there are just two Nash equilibria: the choice of H by both, and the choice of S by both.

In the ten-player version, there is, as we saw, another kind of Nash equilibrium. This happens when two or more people choose H and two or more choose S. They all get nothing, but, as we already saw in the previous section, no one can individually do better. This is a bit like Thomas Hobbes's famous idea about the state of nature: people live completely individualistic lives and no one can individually change this.[8]

The concept of Nash equilibrium has turned out to be a useful tool for thinking about real-life problems, especially those concerning human cooperation and our ability to solve collective-action problems. It also provides tools for analysing the role of conventions and constitutions, and helps us formalize and shed light on philosophical ideas associated with Enlightenment philosophers.

I shall return to these matters in later chapters. What I want to do here is give readers a glimpse of Nash's famous theorem about the existence of equilibrium in games. One nagging question that persisted for a long time was that, while we knew how to define an equilibrium, we could not be sure if an equilibrium existed, especially in large, complicated games. In the simple examples like the ones given above, it is easy to answer this by literally searching through all possible outcomes.

Similarly, we can create simple games where we can easily check that there is no equilibrium. Consider the zero-sum game, where two players each have to choose between A and B. In this game, each player gets either $2 or -$2, so that the sum of their payoffs in each outcome adds up to zero, hence the name. If you win, I lose. Consider now a specific zero-sum game, where, if both players choose A or both choose B, Player 1 gets $1 and Player 2 gets -$1. For all other outcomes, Player 1 gets -$1 and Player 2 gets $1. In this game, it is easy to see that, no matter what the outcome, one player will want to change their position. Hence, this game has no Nash equilibrium.

There are many real-life situations which are, effectively, zero-sum games. When two nations fight for the gold that has just been discovered on a remote island, they are locked in a zero-sum game. One nation's gain is the other nation's loss. However, people often make the mistake of treating all inter-country conflicts as zero-sum games. That is not so. When two countries explore setting up a new factory, if they cooperate they can share a large positive output from the venture, but if they fight they may fail to produce anything. In these non-zero-sum games there is scope for cooperation.

It is easy to spot the equilibria or the absence of an equilibrium in the games discussed above. But reality is complex, with millions of people locked in complicated games of war, diplomacy and financial strategizing. In subjecting these to game-theoretic analysis, it is important to know which of these games have an equilibrium. This question turns out to be extremely difficult to answer. It remained a black box for a long time, for a good reason: we needed a theorem in pure mathematics to solve it. The problem was solved in 1941 by the mathematician Shizuo Kakutani, and his solution is now known as Kakutani's fixed-point theorem. This theorem would, in turn, be used by John Nash to characterize games that are bound to have at least one equilibrium.

There is no need to go into the mathematics here, but we can sample the flavour of the fixed-point argument using a puzzle. Treat this as an exercise for the mind – to use intuitive thinking to solve problems, and to draw our attention to the meaning of what constitutes 'proof'.

A man walks up a mountain path one morning, starting at the bottom at 6 a.m. It is a slow and arduous climb. He stops and starts, walks slowly for some stretches and fast for others, and reaches the peak at 6 p.m.

The next morning, he starts his descent at 6 a.m., walking down the same path, and reaches the same place at the bottom some time in the evening. The question that you have to answer is this. Does there exist a point on the path – call this a 'fixed point' – where the man will be at exactly the same time on both days, the first day on the way up, and the second day on the way down? Whatever your answer, try to back it up with a proof.

Once you have given this some thought and tried your hand at proving your answer, read on to see whether you were right.

Here's the answer: in this problem a fixed point must exist.

This is true no matter how fast or slow the man walks, or how many times he takes a break. And here is a proof. The second day, when the man begins his descent at 6 a.m., imagine a woman begins at the bottom, walking up but *exactly mimicking the speed* at which the man had travelled the previous day. Clearly, the man and the woman will encounter each other on the path at some point. Where they meet is the fixed point. End of proof.

In the end, this turns out to be so simple that one is left wondering if this is really a proof. But it is, because all of us can now see clearly that a fixed point exists. This leads to a fascinating point about proofs. There is no hard definition of what a proof is. Basically, a proof happens when all reasonable people can see this has to be true. Since mathematics and also game theory are disciplines of rigour and precision, such a vague definition of what constitutes 'proof' troubles us. But there is no escape from intuition. In the end, we rely on this much more than we realize.

Why we don't get angry at tigers

Chess, bridge and football are games. The Prisoner's Dilemma, Stag Hunt and nuclear war are games (albeit with very different payoffs). In this sense, we can see life itself as a game – one large, grand game, with billions of players pursuing their own payoffs, strategizing how to make their choices. This grand game that has no bounds except by the laws of physics and biology, which restrict what we can do, is called the 'game of life'.[9] The reason why learning game theory is important is that, while we may not play chess or bridge, we may refuse to play Stag Hunt and we all hope to be lucky enough not to get embroiled in a nuclear war game, no one can escape the game of life. For that reason, learning the elements of game theory is learning to live.

From the abstract discussion above, we have already picked up some insights that can lead us towards a better life. When we describe a game, we are explicit in stating that a player can choose from what is available in his or her feasible set of actions or strategies. That is the only thing that he or she can control.

In game theory we do not discuss people's resentment and anger about what other people do. Other people's actions, including how they may be influenced by you, like the movement of the billiard ball, are an immutable part of life. There is nothing you can do about these 'laws' of nature; and, for that reason, there is no sense in fretting.

This is something we can take away from this feature of game theory for life. Anger and resentment about what others choose or do are, most of the time, pointless emotions. One may not like what others do, but to be angry about what they do is more to hurt oneself than anybody else. Of course, in extensive-form games like chess, we have to think of how others will choose after we have chosen, and so we may strategize our own actions to influence the actions of others. But there is no room for resentment or anger in this.

A 'bad' human being (because, for instance, he has a payoff function in which he gets pleasure from harming others) is like a tiger. Of course, we have to take actions to deter tigers from harming us, but we do not get *angry* at tigers. In fact, we would be much worse at handling them if we became full of anger and resentment when we were attacked. We take that to be a part of life – tigers are the way they are meant to be. We dispassionately think of our best strategy in handling them.

Love and empathy are productive emotions which directly enhance our payoff and well-being, so those should be nurtured and encouraged. But anger, hate and resentment make us worse off and achieve little. This is the *ataraxia* that Pyrrho, Epicurus and the Stoics – from Crates and Hipparchia, through Zeno of Citium, to Seneca and Epictetus – practised as philosophy and as a way of life.

But one does not have to be a Stoic or an Epicurean to practise this. We have seen the spirit of *ataraxia* in philosophers of different persuasions, in the occasional political leader, in theists and atheists. As a young man fighting for India's independence, Sri Aurobindo was held as an undertrial in solitary confinement by the colonial government for a year in Alipore Jail, Kolkata. In one of the most beautiful prison essays ever written, *Tales of Prison Life*, he discusses how solitary confinement can breed hatred, anxiety and even insanity. He suffered all those in the initial days, but overcame them, and the year behind bars turned out to be his year of transformation into the spiritual leader he eventually became. In his own words, 'I have spoken of a year's imprisonment. It would have been more appropriate to speak of a year's living in a forest, in an *ashram* or hermitage.'[10]

We often see in discussions, from policy debates at the highest levels of government and international organizations, to department meetings in universities, how easy it is to lose it. When that happens, the speaker's aim becomes to get their say. If they do this to influence other people's behaviour, there may be some rationale for it. But when one is angry, that is seldom the case. In such situations, getting your say becomes an end in itself. This is a lesson worth taking away from theory to life: try to get your way, not your say.

In giving this advice, I hope that the person's payoff function is a noble one, so that what he or she is trying to achieve is a social good. However, whether or not my hope turns out to be true, the dictum, unfortunately, remains valid.

There is one caveat to the above recommendation. As briefly noted in Chapter 1, at times anger does play a role in influencing the outcome of the game by influencing the behaviour of others. Others may get scared or feel remorse at your anger and change their behaviour. However, even for this, strictly, you do not need anger but the *display* of anger. So anger is never worth it, even though the display of anger might be.

The handling of anger is one specific example of a more general lesson about the management of emotions, and the nurturing of happiness and success. Emotions that rob us of our peace of mind, such as anger and resentment, for sure, but also hatred, jealousy, irritation, bitterness, vengefulness, meanness, despair – it's troubling how long the list can be – are part of the human psyche for a reason. They enable us to take actions that in a normal state we may not be able to take. These actions may have survival value, and so the negative emotions which provoke these actions are there because they have passed the test of natural selection and evolution.

However, if we can train our minds to take the actions that these emotions help us to take, without actually harbouring these negative emotions in our heads, we can get their benefits without their negative side effects.

To do this successfully, we need to train our mind to store the information that gives rise to these negative reactions, not as emotions, but as facts and data. We can then use the information dispassionately for decisions and actions, without troubling our mind. This is not easy but not impossible. If we see a person acting mean to someone else, we usually get upset and take steps to stop this person from being mean. But it is possible to train our mind not to get upset, while still taking the steps that being upset would have prompted us to take, by exercising reason. Indeed, we may be able to take even better steps because the calmness of mind creates clarity of vision. This argument is an endorsement of what many good religious leaders tell us: hate the sin, not the sinner. Further, as we shall see in Chapter 4, in the context of the philosophy of determinism, not hating the sinner is not just a good strategy in the game of life, but it also stands to reason.

The best example of the broad point I am making here comes from our own experience during the COVID-19 pandemic. After years living with the dark clouds of a pandemic over our heads, there has been widespread anxiety and mental health problems around the world. Having to live with a heightened risk of hospitalization and death, and having to decide whether to meet with friends, to travel or even to go to the neighbourhood cafe put a strain on our daily lives. Each decision became a reminder of the dangerous world we were navigating.

Now step back for a moment. If you travel by road from Delhi to Agra, the risk of death is arguably greater than that of dying from COVID-19 after being in situations where you are likely to have been exposed to the virus. But you do not have an anxiety attack when you drive to see the Taj Mahal. This is true even for those who know the data on the risk of accidents, and this is true for those who drive this route regularly, such as taxi drivers.

The reason is simple. We store this data not in the emotional compartment of our brain but in the intellectual compartment.

We take all the necessary steps when we set out. We fasten the seat belt, remind ourselves not to drive above a certain speed, check the brakes regularly and so on. Having taken these steps mechanically, we set out with joy to see the great monuments of Agra. These actions are prompted by rational thinking, not fear and anxiety, which were the instruments many of us used to steer us through the COVID-19 pandemic.

To battle anxieties, we need to take a lesson from drivers on the highway. Take the right steps using reason (not emotion) and, thereafter, live as though there were no risks. Change what you can, then live life like Bertrand Russell's chicken.

3

The Arithmetic of Anxiety

Why everyone is attractive on Miami's South Beach

Self-esteem is an important source of human happiness. To see that others hold us in high esteem builds confidence and can be a source of contentment. Equally, for individuals to believe that people have little regard for them can be a cause for shame, anxiety and even depression.

Economists often make the assumption that human happiness – or 'utility' or 'payoff', as they call it – depends on your level of income and wealth. And that is what people try to maximize. There is indeed some connection between income and wealth, on the one hand, and happiness, utility and payoff, on the other. However, once you have achieved a reasonable degree of economic security, for happiness and life satisfaction, the greater need is for human regard and esteem. This is true of the shy child playing with classmates in school, the outgoing office worker, the socialite and the recluse.

A major source of our unhappiness and disquiet in life stems from a sense of shame. As Martha Nussbaum writes in her book *Hiding from Humanity: Disgust, Shame, and the Law*, 'Like disgust, shame is a ubiquitous emotion in social life . . . Most of us, most of the time, try to appear "normal" . . . Sometimes, however, our "abnormal" weaknesses are uncovered anyway, and then we blush, we cover ourselves, we turn away our eyes. Shame is the painful emotion that responds to that uncovering.'[1]

With shame and lack of self-regard come inferiority complexes. These are surprisingly commonplace. The unemployed or the homeless may have their sense of inadequacy on display. The strutting right-wing vigilante and supremacists of various stripes often have the same complex, though they try to cover it up. Amal Clooney is right when, in introducing the book *How to Stand Up to a Dictator* by Maria Ressa, the winner of the 2021 Nobel Peace Prize, she writes, 'It is ironic that autocratic leaders are often called "strongmen" when in fact they cannot tolerate dissent . . .'[2]

The source of our shame – and autocratic leaders have ample quantities of that – is often hidden deep inside our psyches. Psychologists have studied the roots of some of these pervasive feelings. Martha Nussbaum and others have pointed out how shame has at times been used to curb antisocial behaviour and, as such, forms part of a nation's unwritten laws.[3] But the struggle to avoid shame can result in unsavoury competition. As Erving Goffman (1963, p. 128) observed in his seminal book, *Stigma*: '[Ultimately] there is only one complete, unblushing male in America: a young, married, white, urban, northern, heterosexual Protestant father, of college education, fully employed, of good complexion, weight and height, and a decent record in sports.'

What I want to argue here is that while a lot of our sense of shame may have deep psychological roots, much of it also arises from simple flaws in reasoning. In such cases, we need not Sigmund Freud and Carl Jung to get over our inferiority complexes, but David Hume, Bertrand Russell and John von Neumann to help us reason our way out of them.

I want to be careful not to overstate this. I am of course aware that there are many situations in life when our lack of self-esteem has roots in deep psychoanalytic factors, stress disorders and childhood traumas, and we may well need to have treatment and counselling for it. But, at the same time, there are numerous instances when our inferiority complex stems from a simple flaw in reasoning, which leads us to believe we are in a worse position than we are. A little bit of arithmetic is all we need to break out of it.

Before proceeding with the analysis, it is worth pointing out that inferiority complex is very different from humility. Humility arises from the recognition of our minuteness and relative insignificance in the vast, complex universe. Humility is a desirable quality that has a calming effect and is worth nurturing. Apart from anything else, it has the advantage that others cannot humiliate you because you have done it yourself.

Inferiority complex is emotionally damaging because it amounts to deprecating yourself relative to others. As I will show now, more often than not it arises from an inadequacy of reasoning in our heads. The argument that I will use has a long tradition in economics. It goes back to the 1970s, to George Akerlof's (1970) paper on the 'market for lemons' and Joseph Stiglitz's (1975) work on screening. Without going into the details of these papers, which triggered the rise of a literature on so-called 'asymmetric information' in economics, I want to explain simply why so many people have a sense of inferiority when they should not.

In a nutshell, the reason is this. Most human beings like to display their better selves to the world. The photos they post of themselves on Instagram show them in a good light – looking handsome or pretty (usually handsomer or prettier than they generally are), having fun or enjoying a splendid holiday. Presented only with evidence of other people living their best life, it is easy for observers to believe that their own lives are distinctly worse, and so they begin to withdraw. What is interesting and not commonly understood is that this kind of reasoning has a snowball effect that can end up adversely affecting the vast majority.

Consider a beach in Miami. Suppose (for the sake of our game) that in terms of beauty, the human physique can be classified into eight categories. We can use a scoring system from 1 to 8 to describe these, with 8 denoting the highest score one can get – the most attractive – and 1 the lowest.

Suppose that in Florida there are 1,000 people of each of the eight types. (Yes, I am asking you to pretend that the population of Florida is 8,000.)

Let's assume that in everyday life we do not see one another's bodies, as they are covered up, but those who show up on Miami's South Beach in their swimwear can easily be sized up against one another. Suppose that human beings are reasonable enough not to insist on being the *most* attractive, but they do not like to be in the bottom half. Once they become aware that they are in the bottom half, they feel a sense of shame and develop a complex. Given the above assumptions, half the population of Florida would feel bad about themselves (i.e. the 4,000 people who get scores of 1, 2, 3 and 4), if there was full information.

I must pause here to point out that we really have no reason to feel shame or have a complex for this reason, and I will try to persuade the reader of this in the closing section of this chapter. What I am assuming here is not that this is the way people *should* feel, but that this is the way many do.

Continuing with the thought experiment, assume further that if people feel they are in the bottom half, they prefer not to be seen on South Beach in their swimwear.

I am aware that these are all strong and rather specific assumptions. But they all capture broad features of reality, and we can use them to demonstrate a strange dynamic that plays out in life.

So our beach in Miami opens for the first time on a Sunday. Everyone comes out to the beach. Even though each beach-goer knows what he or she looks like, they do not know where they stand in the overall distribution of attractiveness until they turn up.

By the day's end, types 1 to 4 will realize they belong to the bottom 50 per cent of society in terms of attractiveness. They will develop a complex and decide not to go to the beach again.

On Monday, only groups 5 to 8 turn up. Looking around the golden sands, those on the beach will realize that being of type 5 or 6 is to be in the bottom half, and the latter too will decide to stay away in the future. Remember, people do not insist on being on top, but no one wants to be in the bottom half. They feel ashamed. So on Tuesday, only groups 7 and 8 turn up to the beach.

It must by now be clear that on Wednesday only the 8s bother to turn up.

And this is the equilibrium. The distribution on the beach has stabilized. One thousand people, all of score 8, show up on the beach, creating an illusion of a more beautiful world than is the case. All the type 8s coming out to South Beach and everyone else staying away is the only Nash equilibrium of the Miami Beach game.

Once the equilibrium is reached, for day after day, and year after year, the beach will look like this, with the history of how we got here long having vanished from memory. We will see only type 8s on the beach. An attractive 6 could turn up one day and still feel ashamed. Those peering out of their home or hotel window at the beach may imagine that that is what the average of humanity looks like. All but the top 12.5 per cent of the population will believe they belong to the bottom 50 per cent. Many of them will begin to suffer from a lack of self-esteem.

Some of these people may grasp that, just as they are not going out themselves, there may be others who are staying in. These people will realize that average humanity is not what they see on the beach but includes those who, like them, discreetly stay away. Once you see through this reasoning, you realize that there is a systemic bias in what we see, and through a certain dynamic, the bias is magnified.

Most of us are, in reality, better than we think we are. Facebook, Instagram and Twitter have made this problem even worse. Previously, you had to go to South Beach or visit your local gym to experience this dynamic. Now, sitting in your bedroom or study you can see on social media what people look like or project they look like, what people are up to or, more correctly, what people want to show they are up to. These days, you cannot escape comparison by simply not going to South Beach. So it is even more important now that we exercise reason about what we see of others' lives.

This is also a good occasion for a quick revision of game theory. It is not hard to see that all the type 8s coming out to South Beach and everyone else staying away is the only Nash equilibrium of the above 'game'. Note that this game has 8,000 players. We can think of each player having a choice between B and H, where B denotes going out to the beach and H staying at home. Given the preferences described above, the only outcome where no one can do better by unilaterally changing his or her choice is where all type 8s choose B and everyone else chooses H. It is easy to see that, in this situation, if any non-8 goes to the beach, they will find themselves in the bottom half of those on South Beach and so prefer not to be there.

I shall leave it to the reader to verify that any other group's being on the beach, apart from the group consisting of all the 8s and only the 8s, would not be a Nash equilibrium. That is, if any such group happens to be on the beach, some people would want unilaterally to change their behaviour, either by someone who was not on the beach going to the beach or someone who was on the beach preferring to stay away.

Why do my friends have more friends than I do?

There is another 'statistical' reason for human misery and lack of self-esteem that is getting worse in the age of social media. Most people are discovering that their friends have more friends than they have. This can, at first sight, be disconcerting. How come my friends are more popular than I am?

Research by Johan Ugander, Brian Karrer, Lars Backstrom and Cameron Marlow (2011), using a massive dataset from Facebook, has found that this is a despair widely felt. Their study of 721 million active Facebook users revealed that an average Facebook user has 190 friends. On the other hand, the *friends* of Facebook users have on average 635 friends. They also found that 93 per cent of Facebook users have friends who have more friends than they do (Strogatz, 2012).

The above data should alert you that it must in some sense be normal to have friends who have more friends. Indeed, a paper by the sociologist Scott Feld published in 1991 in the *American Journal of Sociology* showed that it is a mathematical truism that, on average, your friends have more friends than you do. If you feel deflated, you may not need counselling but a quick lesson in arithmetic.

The general claim is as follows. In any society, if all individuals have exactly the same number of friends, then each person will have the same number of friends as their friends have. In *all* other cases, the average person will have *fewer* friends than his or her friends have.

To see this, consider a society with three people, Aaliyah, Brian and Chu Hua. Aaliyah and Brian are friends, and Chu Hua, though she has two names, has no friends.

In this society, clearly each person has on average two-thirds of a friend (thankfully, this is only true statistically). This is because two persons have one friend each and one person has no friends. Now let us turn to how many friends people's friends have. Note that Aaliyah's friends (only Brian) have on average one friend, and Brian's friends also have on average one friend. Chu Hua has no friends and so does not figure in the count. In this society, despite each person having on average two-thirds of a friend, people's *friends* have on average one friend.

Next, suppose that in this society Brian has two friends, Aaliyah and Chu Hua. Aaliyah and Chu Hua have one friend each, Brian (but they aren't friends with each other). On average, people in this society have a bit more than one friend (4/3).

Aaliyah's friends have two friends. Likewise, Chu Hua's friends have two, and the average number of friends that Brian's friends have is one. Thus, on average, people's friends have 5/3 friends. Once again, on average, people's friends have more friends than they do.

With patience, you can check for larger societies and other combinations and networks of friendship that this will always be so. At least some of our seemingly complex social and psychological problems are not social or psychological at all, but a matter of simple maths.

From John Nash to Ogden Nash

The example of Miami's South Beach in the above analysis was not picked out of the blue but from experience. Visiting the beach many years ago with my wife and two children, I remember being struck by the thought that I did not know human beings were so uniformly good-looking with such good physiques. I was so impressed that on returning from Florida, I immediately took to bodybuilding. After one week I decided that, as far as I was concerned, a good physique was not worth the agony of lifting weights every day. So I stopped, and since then I have started and stopped several times. (This is not advice I would give to others, but what I did.)

The example above was based on the assumption that if we are in the bottom segment of some ranking, such as physical beauty on South Beach or how many friends we have, we tend to develop an inferiority complex. The analysis then went on to show that *by this criterion* more people will develop an inferiority complex than should. I want now to persuade you that this criterion itself is wrong. We must learn not to have a complex, no matter where we believe we stand in social rankings.

Human talent and attributes have many dimensions. Even within a single category, such as intelligence, there are many variants. A person with no mathematical skill may be a literary whiz. Someone with little mathematical or literary skill may have business acumen. Someone with little mathematical or literary skill and no business acumen may have a reclusive talent for art. Inferiority complexes arise from getting hooked on treating one skill as special or essential. Luckily, society does not have an agreed way to rank and aggregate these many dimensions. Hence, overall human rankings are destined to be incomplete. You can be low down on one and high up on another, and since we do not know how to weight these different skills, we do not have a way to rank all individuals against one another.

There is a deeper, philosophical matter that we should consider. Rankings are not just social constructs but, in the very end, they are constructs of our own mind. How I see myself in the social rankings is my construction of the social construct. The world around me is, in the final analysis, a construct of *my* mind. And so, often, what I take to be objectively true may not be so. When walking on Miami's South Beach, or thinking about how many friends I have online, the importance of these matters is created by *me*.

Indeed, it is surprisingly hard to prove that other people actually exist. Several philosophers have expressed doubts about the certainty of the existence of other people, since, in the end, whatever we perceive, we perceive through our minds. I cannot rule out (and I do not rule out) that others may be a construct of my mind. I was both troubled and relieved when in my high-school years in Kolkata I reached this conclusion on the basis of my own introspection. It was comforting later to read that Bertrand Russell had to deal with this too. Russell went on to argue that, on balance, it was probably the case that others do exist outside our heads. I think the thought of a world with no one besides his own mind was so troubling that he persuaded himself that he was *probably* wrong.

My own view is that since my mind perceives others so fully and I have the capacity to see them, love them, hate them, chat with them, that is good enough reason not to feel lonely. On whether they actually exist, my hunch is that they do, and of course I live as if they do, but I am aware that there is little basis to that hunch. Also, while our hunches and common sense are generally reliable, they do sometimes fail. For thousands of years, people took the Earth to be flat not by using hard logic but by using their hunches and common sense – after all, it's generally flat enough where we're standing. So, while we live life as though others exist independently of our own minds, a subliminal awareness that the world I perceive is a creation of my mind can be a source of comfort and strength.

The philosopher who popularized this line of thinking was René Descartes, who famously said, '*Cogito, ergo sum*' ('I think, therefore I am'). It is a philosophically powerful but, at the same time, potentially dangerous line of discourse that can lead some people to solipsism. Let me therefore end this with the best attempt at a rebuttal of Descartes that I have read, even if it is a bit frivolous. This also brings me back full circle to Nash, though this time to Ogden, not John. To René Descartes's observation, 'I think, therefore I am,' the poet's response was – I am paraphrasing here – *most people do not think but nevertheless they are*. As Nash puts it:

> Descartes was one of the few who think, therefore
> they are,
> Because those who don't think, but are anyhow,
> outnumber them by far.[4]

In closing, a reminder from Ogden Nash's life of how varied human talent can be. Ogden Nash went to Harvard in 1920 but dropped out after a year. It didn't quite work out. Thereafter, he and his family wanted him to be a bond trader. Luckily for future generations of readers, Ogden Nash showed a total lack of talent for this, too. In his own words: 'Came to New York to make my fortune as a bond salesman and in two years sold one bond – to my godmother.'

4

Scepticism and Paradox

God, scepticism and a way of life

Life is full of the unknown. Being conscious of this is not just a matter of modesty, but of intelligence. The folly of overconfidence has been the cause of some of the biggest catastrophes, for individuals, groups and even nations. One sees this mistake in theists and in atheists, in believers in science or in superstition.

As an atheist, I live on a rather meagre diet of beliefs. The only thing certain is how little we know. This ignorance must extend to what lies beyond what we perceive with our sense organs: the eyes, mouth, nose, skin and ears.

Let me explain. I grew up in a traditional Bengali household in Kolkata. My mother used to take great pride in the fact that I spent a large amount of time as a child praying and meditating (no doubt with my mother's encouragement). Like all traditional people, we went regularly to places of worship. I loved the visits to the Dakshineswar and Belur Math temples on the banks of the river Ganga, at the edge of the city of Kolkata. These places were beautiful monuments to the mystery of the universe, and also to the history of Kolkata and spiritually inspired individuals like Swami Vivekananda, who made enormous personal sacrifices to travel the world and spread his message of universal brotherhood.

Later, when I taught at the Delhi School of Economics, I got to meet Mother Teresa in Kolkata, through my sister who worked with her. I heard from my sister of Mother's endless compassion, of how she would not have a moment's hesitation in hugging a dying beggar or helping a leprosy-ridden street dweller to find a home. It is likely that I would have had disagreements with Mother Teresa about her views on God, religion and how modern medicine works, but meeting her in her nondescript home, with the low-voltage evening lights coming on in the neighbouring houses, it was impossible not to be moved and marvel at the power of human kindness.

The ambiance of places of worship can indeed be arresting. I loved and still love the rituals associated with religion. It is the choreography of rituals that appeals to me. Many progressive thinkers say that they believe in religion but reject rituals. My own position is almost the opposite of this: I do not believe in religion but I love the rituals.

I lost my belief in God some time in high school, maybe at the age of thirteen or fourteen. There were several factors. Through my childhood, I remember my father going to the temple of Goddess Kali every Tuesday evening. I heard him, on several occasions, mutter as he got into the car to go to the temple how he did not believe in God, but did not want to take any chances. Much later, I learned this was similar to the line taken by the seventeenth-century French philosopher Blaise Pascal – the so-called Pascal's wager. In retrospect, I don't think my father was a believer. Listening to him, and thinking for myself, my belief had begun to waiver.

Then came my discovery in high school of the writings of Bertrand Russell. This happened through a friend, who was one or two years my senior. He was an extremely well-read person, and in the course of our many conversations he insisted I should read Russell.

I wasn't a big reader, but I thought I would try reading Russell. This turned out to be one of my most important intellectual experiences in life. What I took away from Russell was not that what he said was right, but his dictum that you should never accept anything just because someone or some book has said it. I loved his books, and guzzled his *History of Western Philosophy*.

Bertrand Russell helped bring to the fore what was, in a somewhat inchoate way, already beginning to form in my head. I decided I did not believe in God. Among erudite Hindus there has long been a debate about mono- and polytheism. Is there one God, or are there many? The conventional Hindu belief in many Gods was, even to many traditional Hindus, unacceptable. Some argued that the many Gods were meant to be manifestations of the one divine God.

I did not find this debate about there being *n* Gods, and whether the *n* should be one (monotheism) or greater than one (polytheism) interesting. As long as the *n* was non-zero, it seemed to be wrong. The belief that there exists a God who is both all-powerful and all-merciful, as virtually all religions believe, is inconsistent with the state of the world around us, with its great tragedies and sorrows. The fact that I am occasionally sad makes this definition of God illogical, because

God's being all-powerful and all-merciful, and my being sad, are together impossible.

What about other views of God – for instance, someone who is fairly powerful and somewhat merciful? Could such a creature not exist? The answer is: yes, but I would not call that creature God.

What about God as the creator of the universe, someone who deliberately created everything? That is a logical possibility. We cannot firmly rule out such a God. On this, I rely, admittedly, on hunch. I do not *think* the universe was created by anybody deliberately. Further, even if it were, why should that agent be worshipped? However, concerning the existence of a creator, I keep myself open to the possibility that I am wrong. That is the essence of scepticism, maybe the school of philosophy closest to my heart.

Let me recount a remarkable incident in connection with a trip to Turin, Italy.[1] In 2017, I went there to deliver the 15th Luca d'Agliano Lecture. It was a memorable visit. Turin is the city where the Marxist philosopher and politician Antonio Gramsci studied and started the weekly newspaper *L'Ordine Nuovo*, and took to political activism before being arrested and jailed by Mussolini's police. It is the city that Nietzsche lived in and loved, where he had his famous mental breakdown and eventually lost his sanity.

It was not just a magical visit but also miraculous. A week before I travelled, my US Green Card had gone missing. I turned our home in Ithaca upside down, but to no avail. If I did not find it, I would have no choice but to cancel my trip. I would not know how to break such news to the organizers. I joked with my wife that it was time to try prayer, and sat down self-consciously, legs crossed like a monk, and prayed: *God, as you well know, I don't call on you every day. In fact, I pray once every several years only when I am desperate; and today is one of those days, since it will be truly embarrassing for me to cancel this long-planned lecture. I doubt that you exist but, if you do, please appreciate my honesty and give me my Green Card. At the same time, I am not saying that if you do, I will become a believer. Looking around the world, there is so little evidence of your existence that one miracle is unlikely to make me change my mind.*

I then got up, did my usual late-night reading and writing, and went off to sleep. Next morning, unmindfully, I opened the drawer next to my bed which I use all the time and which I had, I seemed to recall, searched carefully. And there it was, in plain sight: the Green Card. I felt an emotional charge, a mixture of elation and confusion. I tried to reconstruct all the events of the previous few days and could in no way that would traditionally be called 'scientific' explain what had happened.

So what do I make of it? My inclination still is to believe that the Green Card was always there and neither of us saw it. On the other hand, violations of the laws of induction do not trouble me because of my innate scepticism. I believe that whatever is not *logically* impossible is possible. However, what I experienced was so confounding that the only way I found to sum it up was: whether or not God exists, He loves me.

The following year, when I was visiting Kolkata, I told my eldest sister about this incident. She shuddered and said, 'I have had it. No wonder God does not listen to me. I disturb him by praying every day.'

Here is the lesson from this. A miracle is not a *logical*

impossibility. It is an event that jars our sense of the way the world is. Passports do not typically vanish and then reappear after the owner prays. Nevertheless, we have to be open to such events. This is where Bertrand Russell's observation, cited earlier in the book, about the naivety of the chicken that gets a jolt one morning when the man who cared for it every day wrings its neck instead, becomes relevant.

Science cannot save us from this predicament. This is the reason why, although I share Richard Dawkins's atheism, I find him too certain, and his dogmatic belief in the power of science not that different from superstition.

The philosophy I am appealing to has a long history. It goes back to Pyrrho, his philosophy and his way of life. Pyrrho, born in Elis in 360 BCE, lived – especially correcting for the shorter life expectancy of those times – a very long life, ninety years, no doubt helped by the calmness of mind that he achieved through his philosophy.

Pyrrho wrote nothing. What we know about him is all hearsay, but he had such an enormous impact on his followers that his life and sayings are still reasonably well documented. He was the ultimate practitioner of scepticism, the school of thought that held doubt as the essence of life. Stories abound about Pyrrho. It seems that once, Anaxarchus, twenty years his senior and also a believer in scepticism, fell into a ditch while out walking. Pyrrho, walking past, saw him but made no effort to help him out of the ditch. He later explained that he could not be sure Anaxarchus would be better off outside the ditch than in it.

They remained friends. Both of them travelled with Alexander's army to India and it is believed that Pyrrho was impressed on meeting some *mouni sadhus* in India. *Mouni sadhus* were not only like Pyrrho in not *writing* anything, but they did not *say* anything either. They just stood in silence (the meaning of *mouni*), which Pyrrho admitted was superior. Another occasion for defeat was on a sea voyage, when a big storm broke out. Those on the ship were terrified as they were tossed about by the waves. There were just two beings on board who were calm – Pyrrho, and a pig. However, Pyrrho could not get himself to do what the pig did through the storm: keep eating.

One has to be careful in taking lessons from scepticism. As one would expect, there are few. Indeed, while scepticism led Pyrrho to write nothing, Bertrand Russell, an avowed sceptic, wrote prodigiously. Pyrrhonism has few prescriptions beyond nudging us to question what we learn and believe we know, and to keep our minds open to different explanations and perspectives. The hope is that this attitude will help us deal with the challenges of life with equanimity.

Assumptions in the woodwork

A valuable by-product of scepticism is that, by advocating questioning, it has acted as a major impetus for science and resulted in some of the biggest breakthroughs in the world of ideas. In the practice of science, as in everyday life, we make assumptions. Indeed, we have to make assumptions as we proceed. The trouble stems from the fact that some of those assumptions are so ingrained in us that we are unaware of them. They then become a part of the woodwork of the discipline.[2]

Augustin Cournot, in 1838, created a beautiful model of a market with a limited number of competing producers. This gave us a grasp on how real markets work. The trouble is, with the routine use of this model in analysing markets and drafting laws and regulations, we tend to forget the many assumptions Cournot made to build the model. Then, when reality fails to live up to the model, as we saw happen with the malfunctioning of the vaccine market during the COVID-19 pandemic, economists get surprised.

There is an interesting connection here between science, magic and miracle. The reason why magic or a miracle surprises us is that we make implicit assumptions in our head when we watch magic shows. When we see the rabbit being pulled out of the hat that seemed to be empty a moment ago, we get a surprise because we make false assumptions about the ways in which a rabbit could have got into a hat. When the magic is explained to us, the feeling is: *Of course, I should have realized this. It is so obvious.*

In economics, as in other sciences, we try to make our assumptions transparent by writing them down explicitly, often calling them axioms. In the practice of normal science, we question these explicitly stated assumptions, and try to collect empirical evidence to check whether they are valid. If they seem invalid, we change these assumptions and try to reconstruct some of our models. These exercises can lead to tinkering and improvements. But the biggest breakthroughs occur when, on those rare occasions, someone realizes that the flaw lies in an assumption of which we were not even aware, because it is such a part of the discipline's woodwork. It is the sceptical mind that is most likely to stumble upon these discoveries because of its propensity to question. For the same reason, the sceptical mind also helps us avoid mistakes in everyday life and enables us to lead better lives.

One of the most famous discoveries of an assumption in the woodwork happened in geometry. Euclid wrote down a model of geometry, which in terms of elegance and beauty had few parallels. The entire discipline was built up from scratch, from axioms to the theorems that emerged out of them like magic. There was, however, (at least) one assumption that Euclid made unawares, because it seemed so natural, and he never wrote it down as an axiom. His assumption was that the entire exercise was being done on a plane of two dimensions, like a tabletop. This was realized much later.

The discovery of this did not need a mathematician. It needed an individual with the ability to question what *seemed* obvious. Among the first to realize that there was this assumption in the woodwork of geometry was Omar Khayyam, Persian poet, philosopher, historian and mathematician from eleventh-century Nishapur. Khayyam was followed by Ferdinand Karl Schweikart, German jurist, who did mathematics as a hobby and stumbled upon Euclid's hidden assumption in the early nineteenth century. Of course, the reconstruction of a new 'non-Euclidean' geometry would not have been possible without mathematicians, and we have Nikolai Lobachevsky, Friedrich Gauss, Bernhard Riemann and others to whom to be thankful for this.

This discovery had serious real-world implications. Most importantly, some results of Euclidean geometry do not apply to the surface of the Earth, since that is not a plane but approximates to the segment of a sphere. If we had continued to use Euclidean geometry into the age of rapid travel across the world, making calculations of how planes would fly, we would likely have made calamitous mistakes with lethal consequences.

The use of axiomatic method is also now common in economics and reached a pinnacle with Gérard Debreu's (1959) slim book, *Theory of Value*. The book is an attempt to capture the breadth and magnificence of Adam Smith's, and later Léon Walras's, conception of the entire economy with mathematical precision. His book is written up entirely in axioms and theorems. He did for economics what Euclid did for geometry. It is a book of remarkable elegance, like poetry. This and the work of Kenneth Arrow had a huge influence on economics because they formalized Adam Smith's theorem of the invisible hand of the market, which helps coordinate the actions of selfish individuals and enables them to achieve the collective good, unwittingly.

This mathematical formalization was important because, by writing down several assumptions explicitly, it drew to the attention of the *astute* observer the assumptions needed for the invisible hand to work but not written down explicitly. Arrow was cautious enough to remind us of this on several occasions (see Arrow, 1978).[3] There are assumptions that are so deeply embedded into the structure of economics that we forget they are needed for the invisible hand to work. For instance, because these are written down explicitly, economists are clear that people's preferences need to satisfy some technical properties like transitivity. Transitivity is the assumption that if a person prefers apples to bananas, and prefers bananas to oranges, then the person must prefer apples to oranges.[4] It's usually true in life, but not always. Likewise, we often assume that human preferences satisfy the law of diminishing marginal utility; that is, as we consume more of the same good, the utility that we get from each additional unit of the good goes down.

Textbook economics argues that if these assumptions are valid, then there will be trade and exchange among people. As a result, most students of economics believe it too. What throws a spanner into this theory is that these assumptions hold true for rats as well. However, rats do not perform trade and exchange with one another. This is a reminder that the above assumptions may be *necessary* but they are not *sufficient* for trade and exchange to occur.[5]

Many of the hidden assumptions that we need for trade and exchange to occur and for us to achieve the collective good have to do with the fact that an economy is embedded in society, politics, culture and institutions. Thus, for instance, for trade to occur, we must restrain ourselves from rushing and snatching all the food from the trader in the marketplace. This is an ability that rats do not have. This restraint has to do with culture and norms. Likewise, we clearly need to talk and communicate in order to be able to conduct trade. However, we do not, among the various axioms that we so carefully record in our textbooks, write down the axiom: *Can talk*.

As the opening lines of Christiansen and Chater's (2022, p. 1) book on the 'language game' remind us, this is true of life: 'Language is essential to what it means to be human, yet we rarely give it a second thought. We discover just how central it is to every aspect of our lives only when it fails us – whether in a foreign city or following a stroke.' Economics is no exception. That we can talk is taken for granted. Often, that is fine, because this ability and other norms and cultural traits needed for economic functioning are usually quite stable. However, certain norms can change over time.[6] It is possible that the digital age is changing the way we talk and communicate with one another. That assumption, *can talk*, may be changing, because the meanings of the same words on social media could be changing, shaking up the foundations of the modern economy and society.

In making sense of the world, to think of novel solutions to the world's problems and even to take better decisions in everyday life, sometimes our answer lies in making visible the hidden assumptions that gird our theories about the world.

Seeing, hearing and mearing

Most human beings are aware, at least subliminally, that there are many realities that are hidden from our sight and consciousness. This awareness is the motivation behind not just science fiction, but science, philosophy and speculation about the existence of extraterrestrial life.

It is fascinating to think that, in some other planetary system, on one planet there may exist intelligent life. The way the seafarers of Europe of the fifteenth century imagined distant lands, peoples and customs, and eventually stumbled upon these weird people in the Americas, India (ancestors of the author), East Asia and elsewhere, we too love to think of people on distant planets, their lives, manners and cultures. How will they react to us when we meet? Will they be kind and communicative, or feel threatened and belligerent and take to war? Of course, much will depend on our intelligence disparities. If the new 'humans' we discover are vastly more intelligent than us, we will be outwitted and our fate will depend on their empathy or ruthlessness. It will be the other way around if we have the relative advantage.

There are, however, other variants of life that we find harder to grasp. It is hard to imagine (but true) that size is endless both in terms of largeness and – what we find more difficult to accept – smallness. My scientist friends tell me that smallness has limits – molecules, atoms, protons, photons, and not much further – and that some of these tiny particles are elementary particles; that is, they are not composed of any other sub-particles. I find it difficult to believe we can ever prove anything to be not composed of anything else. This is because smallness is endless. You do not have to be a scientist to see this. In fact, being a scientist may be a handicap, because there are assumptions hard-wired into your thinking. The need here is not for normal science but for imagination.

Let me explain with a geometric analogy. Consider a universe which consists of nothing but a set of concentric circles. Each circle has two adjacent circles, one having a radius which is double the radius of this circle, and the other with a radius that is half of this one. It follows from this definition that we have an infinite number of circles. Another way to think of this universe is to think of a point that acts as a centre. Around it there is an infinite number of circles (and, in fact, nothing else). Thus, there could be a circle with a 1-mile radius, and circles with a ½-mile radius and a 2-mile radius, and another two circles with a ¼-mile radius and a 4-mile radius, and another two with a ⅛-mile radius and an 8-mile radius. And so on, endlessly.

What is fascinating about this universe is that, in an important sense, any pair – a circle, *x*, and the universe – is identical to any other pair – a circle, *y*, and the universe. For us, observers of this universe stuck at a certain scale, some circles are tiny and some gigantic. But since this universe has nothing but these circles, we observers do not exist. Hence, any circle's position in the universe is like any other circle's position in the universe. There exists an endless number of circles both larger and smaller.

This analogy alerts us to the possibility that the molecules and atoms or some other similar tiny particles in our world could be like tiny solar systems populated with human beings, like us, with the same kind of intelligence, families, friends, cars, planes, wars, love and friendship. We and our solar system, in turn, could be like atoms and molecules, in the toe of a person just like us (except for size), in a universe just like ours (except for size). These are all examples of extraterrestrial life very different from what we often imagine or look for.

There is another assumption hidden in the woodwork of our conception of the world. Consider how we perceive and create the universe in our mind. With our eyes, we see the world – the blue skies, high-rise buildings reaching out to them, the seas stretching to the horizon, flowers in bloom. With our ears, we hear words of love and hate from friends and enemies, political leaders speaking from the podium to no one in particular, the horn of a lone ship returning to the docks at day's end. With our hands, we touch and get a sense of shape and texture. With our nose, we smell. With our tongue, we taste.

For each of these dimensions of the world – the visual, the

melodious or the pungent – we have an organ: the eye, the ear or the nose. There is, however, no reason to assume that the number of dimensions of the world is the same as the number of organs we have with which to perceive these dimensions. It is possible that, just like sight, sound and smell, there is another dimension all around us, but we do not have the organ to perceive and be aware of it. We are able to see and hear but we cannot *mear*, because while we have eyes and ears we do not have *mears*.

To digress briefly here, we must not assume that not having a sensory organ is always a loss. When I was growing up in a large and loud joint family in Kolkata, and an uncle suddenly lost his hearing, many relatives came to console him. As they grieved that the uncle would no longer be able to hear, an elderly aunt, known for her sagacity, sat in a corner muttering, 'Lucky, lucky.'

However, good or bad, the absence of a sensory organ can be of great consequence. To understand what a huge effect this can have on our perceptions, we have to do a mental experiment. Consider someone born blind – in fact, born without eyes. This person may learn to listen and talk, play games, read Braille. Over time, the person may talk about the sky, the stars and the moon, but clearly the person has a very different idea of what these are to others. The sky, the stars and the moon will likely conjure up something different in their head than in ours.

Now go a step further and imagine a world where *no one* has eyes. That organ is just not there in humans. Apart from that, the world is just as it is now. People would have no idea that all around them there is the visual world, which is so different from the dimensions of the world that they are used to, like sound and smell. This huge piece of reality will be beyond their perception and imagination. What I am asserting here is that we may all, unwittingly, be in a similar predicament.

So that is the kind of additional dimension of the world that may be all around us. It is totally different from sound, smell and sight, which we know only because we are capable of hearing, smelling and seeing. But, not having mears, we have no idea what mearing is all about, what being able to mear would reveal happening all around us. If one person arrived on Earth who happened to have mears, like the arrival of a person with eyes in a world of the born-blind, and tried to make us aware of this additional dimension of the world all around us, it would be impossible for us to understand. We would probably ridicule this person or think they were mad. It is, in fact, not impossible that some of the mystics through history, or even mad men in authority who heard voices in the air, may have had this additional sense and were able to mear. A lot of what they said will naturally sound like gibberish to us.

To appreciate the mysticism of life, or at least recognize its possibility, we need to acknowledge the fact that there may be other dimensions of experience. They may be literally all around us, as all-encompassing as the visual world or the tactile world. We are simply unaware of these dimensions.

Can it be rational to be irrational?

Paradoxes are a great way to become aware of our hidden assumptions. There are, I believe, no real paradoxes in the world. Paradoxes stem from our incapacity to think clearly and reason with acuity. An illustration of this is the celebrated Russell's paradox, which Bertrand Russell stumbled upon in 1901, while working on his book *Principles of Mathematics*.

Let X be the set of everything. So a spoon is an element of X, as is the set of all spoons. Now consider a set Y, belonging to X, which is defined as a collection of every set that does not contain itself as an element. Consider now the set of all spoons. This is clearly an element of Y, since the set of all spoons is not a spoon. Now consider the question: is Y an element of Y?

If it is an element of Y, then Y cannot be an element of Y (by definition). And if Y is not an element of Y, then Y *must* be an element of Y (again by definition). Hence, Y is neither an element of itself, nor *not* an element of itself. That is impossible. So what we have is the paradox that Russell had stumbled upon.

Russell's paradox lived with us for a long time. It was eventually realized that it stems from an assumption that we make and Russell was making, unwittingly. This is the assumption that there exists a 'universal set' – that is, a set of everything. Once we realize that there is no such thing as the set of everything, this paradox is resolved because the set Y, as defined above with reference to the set of everything, may not be a set at all.

I will return to the subject of universal sets later, but let me here introduce the reader to a game I developed called the Traveller's Dilemma, which illustrates some paradoxes of game theory.

Two travellers have returned home from a Pacific island, each having bought an identical village artefact from the same vendor. On arrival at their destination they find the artefacts have been damaged, and so they ask for compensation from the airline manager. The airline manager agrees, but says his problem is that he has no idea what the cost of this strange object is. So he proposes the following. Each traveller has to write down a number from 2 to 100 on a piece of paper. If both write the same number, both of them will get that number in dollars. If they write different numbers, the manager will assume the lower number is the correct price, and he will pay the travellers as follows. The one who wrote the lower number will get the lower number in dollars plus $2, as a reward for honesty. The other traveller will get the lower number in dollars minus $2, as a penalty for trying to con the airline.

Taking this payment system as given, the two travellers each have to choose, on their own, a number. This is the Traveller's Dilemma. We assume each person is 'rational', meaning they want to maximize their own dollar income. Further, it is assumed (as almost always in game theory) that all players know that all players are rational, all know that all know that all are rational, and so on. This is referred to as the assumption of common knowledge.

What will be the equilibrium outcome in the Traveller's Dilemma? The reader can apply the concept of Nash equilibrium described in Chapter 2 to find the answer. But let me take a more intuitive route. Suppose you are one of the travellers and your first thought is to write 100, and you feel the other player will think the same way and then you will each get $100. But then it should strike you that if you write 99 instead, you will get $101. However, if that is the rational thing for you to do, then it is rational for the other player as well. But if *both* of you write 99, you will both get $99. So then you can do better by writing 98. You will then get $100. But the other person, being rational, will surely think of that, and so on.

This backward induction continues unabated. The only place where it stops is at 2. So the two rational players will both end up writing 2 and they will get $2 each. That is indeed the Nash equilibrium. The outcome (2, 2) will be the prediction of game theory. But surely there is something wrong with this conclusion?

Laboratory experiments confirm our intuition. The players seldom choose 2. In fact, the most common choice is a number in the upper 90s. The divergence of actual behaviour from what game theory predicts has many explanations – not all human beings are so ruthlessly selfish, not all reason well, and so on. These are interesting findings, but there is an open question beyond these that has not been adequately investigated. It is arguable that even if people were ruthlessly selfish, keen on nothing but maximizing their income, and capable of flawless reasoning, they still would not choose 2. They will roughly reason as follows: if rationality leads us to such a bad outcome, it is rational for me to reject rationality and choose a high number. Surely the same holds for the other player, and both players can see that this is true for both players.

The assertion that it is rational not to be rational is a paradoxical one. It is very difficult to formalize this, and I have not seen any successful effort to do so. Efforts to 'solve' the Traveller's Dilemma have brought in ideas from laboratory experiments and psychology, and used formal arguments from game theory and computer science. It has pointed out and analysed that human beings are not altogether selfish, that rationality is in reality not common knowledge, that our capacity for cogitation is limited, that we may be more keen on avoiding regret than maximizing payoffs, and even how outcomes will change as AI takes over from us in the playing of such games.[7] But the paradox in the reasoning remains unresolved.

One reason I bring this up here is the possibility that the lay reader may have an advantage over the expert in solving it, because her mind is not pre-loaded with assumptions. There is often an incapacity that comes with excessive training. There are primal, philosophical problems that underlie some of these paradoxes which may be more visible to an 'untrained' mind.

Among other things, the Traveller's Dilemma casts a shadow over the assumption commonly made in game theory, namely, that rationality is common knowledge: not only are players rational but they know that they are rational and know that they know that they are rational, and so on. The Traveller's Dilemma makes us wonder if this assumption may itself be flawed, like the assumption of there existing a universal set, which was the source of Russell's paradox.

Is politics too much like football?

One major challenge in taking ideas from game theory to decisions in life is that one has to remember that theories are artificial constructs. We have to use them in conjunction with common sense and reasoned intuition when we take these ideas out into the streets. One assumption in the theory of games which is likely to have many exceptions in reality is the statement that each player has a well-defined set of feasible actions or strategies. In parlour games, like chess and Hex, this is true. In chess, at every stage there is a well-defined set of all possible actions from which the player has to choose. But in everyday life, that is rarely the case.

When game theorists describe the Cuban Missile Crisis of 1962 as a game between Russia and the United States, or the current showdown involving Ukraine, Russia and the United States as a game, they state the options open to each agent. But, in reality, do we really know what the set of actions is? If we read the secret recordings of the conversations in the Oval Room of the White House in 1962, as John F. Kennedy and his advisers discussed how to react to the Russian missiles placed in Cuba, every now and then a new course of action would appear on the table. *What if we do X, Y or Z?* When the final decision was made to blockade the seas around Cuba and give an ultimatum to Russia to start dismantling the missiles, it was not necessarily the best course of action, but it may have been the best course of action among those that were believed to be available.

In most games in life, there are many more strategies available than those from which the player chooses. The game of life refers to the ultimate game: all the players are specified; for each player, all actions that don't violate the laws of physics are treated as available. Thus, when we are at war, or corporations are playing some financial game with others, we are part of the game of life. We can, in principle, do whatever we want from a vast collection of potential actions. But the very idea of the game of life is contentious in the same way that there being a universal set of everything is contentious. In the game of life, the actions open to each player is supposed to be the vast set of 'everything possible within the limits of the laws of physics'. It is possible to argue that this unspecified universal set of everything is not a set at all, in the same sense that the universal set of everything, we now know, is not a set.

The way out of this is to treat the game of life as an agreement among game theorists that in discussing a problem, they will assume that it is the game of life. Thus, if we want to think of the Prisoner's Dilemma, discussed in Chapter 2, as the game of life, we are assuming that the universe has no one else but the two players and each player has no other choice apart from C or D. It is the same with the Traveller's Dilemma. If this is the game of life, then all we have in the universe are these two players, and all they can do is to choose a number from 2 to 100. There is nothing more to life.

The game of life does exist, but it is purely a convention, an agreement between the writer and the reader. Note that this means that, on seeing the disastrous outcome in the Prisoner's Dilemma, we cannot say that we should try to solve the problem by imposing a fine on a player who makes the wrong choice. If there was an agent who could observe how the players behave and fine the ones who choose D, that agent should have been part of the game to start with. If the Prisoner's Dilemma is the game of life, we cannot go beyond the two players of the game and produce new players out of thin air to solve the bad outcome of the Prisoner's Dilemma.

At this point I will simply ask readers to tuck this away in their minds. We will need to think more about it in the next chapter, when we think about morality and how to make the world a better place. As a prelude, I want to discuss one more assumption of game theory which does not pass the test of reality.

Game theory – and, in fact, much of economics – treats human preferences as given. Some prefer oranges to apples, some apples to oranges; some get joy from listening to music, some from seeing art. It is good that economics and game theory are non-judgemental on this. Whatever it is that you seek is part of your payoff function. Recall that in Chapter 2 we described the payoff function as one of the three givens that describe a game. We, the analysts, treat this as a given ('exogenous') and go on to analyse behaviour, outcomes and equilibria.

What we do not recognize in game theory but is pervasive in life is the phenomenon of what may be called 'created targets'.[8] Some striking examples come from sports. Think of football. You construct two rectangular bar arrangements at two ends of a field, give a ball to a group of people and tell them that those wearing red shirts should kick the ball through the rectangle of bars on the northern end of the field, and the ones wearing blue should kick it through the bars on the south, and that you will keep count of which side manages to score most often and declare that side the winner. Soon you will have people falling over one another, willing to take injuries, to get the ball into one rectangle and to stop it from going into the other. You do not need to give the players money or apples, or oranges or clothes. For them, the joy of getting the ball into the rectangle and to watch the score is enough. Once the game gets going, winning becomes its own motivation.

Indeed, you can, over time, get onlookers identifying with Team A or Team B and cheering to see them score, so much so that many will try to get away from work and leisure to watch this game and even be willing to pay to boost the chances of the team they identify with, not to mention beating up supporters of the other group.

This is just one example. Life is full of such created targets, or payoff functions that morph or even emerge during a game. It is not an immutable given. This has big implications for how society functions. Worryingly, electoral politics is often like that. Once people begin to back the Democrats or Republicans in the United States, or the United Kingdom's Tory or Labour party, or Congress or BJP in India, after some time it becomes like people supporting Manchester United or Arsenal. It becomes a created target, an end in itself. It is no longer the manifesto or ideology of the party that prompts support. One backs Republicans because the joy of seeing Republicans win is the same as the joy of seeing Arsenal score a goal in a football match.

For big corporations and powerful organizations, and even individuals like political leaders, this human propensity to adopt new targets and then make choices – even be willing to pay a price – in order to achieve that target creates big opportunities. We see all around us examples of politicians creating new games and targets, which hapless individuals struggle to achieve, and the ultimate winners are the politicians. This can be good or bad, depending on the payoffs that the politicians themselves are trying to maximize. If they are trying to maximize human welfare, then this tactical use of target creation can be useful. If they are trying to line their pockets, then it is not.

Once a target gets into people's heads, that can become an end in itself, and shifting the goalposts around can result in huge shifts in societal outcomes. When I worked at the World Bank, I was responsible for the division that produced the 'Ease of Doing Business' rankings across countries. It soon became evident that for many countries, moving up this ranking ladder had become an end in itself. The measure had become a target. Some governments wanted to move up the Doing Business ladder not to spark greater growth or higher standards of living or less poverty or more jobs, but simply to move up the ranking ladder. That was the game, like scoring a goal in football. This made me acutely aware that these rankings can be misused. If we changed the criteria so that if a country allowed investment banks more space to exploit customers, the country would move up in the ranking, we could have nations creating this extra space for exploitation.[9]

Determinism and choice

I became a determinist when I was a student at Delhi's St. Stephen's College, much before I knew there was a term called determinism, and a whole school of thought built around it.

I stumbled upon this, while thinking about the world, through discussions with one of my classmates, Niloy Dutta. Niloy was a phenomenally intelligent student, the most brilliant in our class, and we spent a lot of time arguing and agreeing. Niloy ended up having an eventful life. He gradually grew disillusioned with the injustices and inequities in the world that we could see all around us. He joined the Communist movement, the Naxalites, and dropped out of college to help bring about the revolution. The Communist movement peaked during my years in college. Eighteen Stephanians left college before finishing their studies, walking out in the quiet of the night to avoid being arrested by the police. Niloy, like the others, would never return to St. Stephen's. He spent a long time in jail. The revolution never happened. He returned to civilian life, went to law college in Guwahati and eventually became a successful lawyer.

I agreed with him and some of my closest friends in college, who were all members of the Communist Party, about the injustices that prevailed in Indian society, but I refused to join the party or join the revolution. Of course, this may have been out of selfishness on my part. I certainly did not want to jeopardize my own comfortable life. I do not think that was the reason, though. Niloy was much more intelligent than me, but I feel I reasoned better than him. I had two problems with Communism. I agreed that the utopian world my friends like Niloy – and, for that matter, Karl Marx, Friedrich Engels and, later, Leon Trotsky and M. N. Roy – visualized was a wonderful world. Marx's ideal of a society where each of us works according to our ability and gets paid according to our need is a wondrous world, one to which we should aspire.

What I did not see was a blueprint of a path that would get us to that ideal. The failure of Communism to reach any kind of utopia, and its tendency to end up creating the opposite, shows my apprehension in college was not misplaced.

The period spent in jail did a lot of damage to Niloy's health. He died of natural causes, but prematurely, in 2021. My decision not to join the revolutionary movement had greatly disappointed Niloy. I was grateful to him that, despite this, we remained close friends, even though we had no contact during his years living underground since he was on the watch list of the police and had to be in hiding. I was also grateful to him for his staggering intellect, our many debates and arguments in college, which led me into a way of thought – namely, determinism – that would be important for me.

The first step to determinism is causality. Causality, like most axioms, cannot be proved or disproved. Causality basically means the following. Suppose there is nothing in the universe but just a string of sounds. Suppose that up to now there has been nothing but an infinite history of equally spaced *tick, tick, tick*... Causality says that in the next moment, the sound cannot be *tock*. If an infinity of ticks resulted in *tick*, then an infinity of ticks must always result in *tick*. To put this more generally, if everything that happened before Moment 1 was the same as everything that happened before Moment 2, then what happens in Moment 1 must be identical to what happens in Moment 2.

Since the existence of causality cannot be (or at least has not been) established deductively, it may or may not be true. Still, while it is not possible to prove causality, most of us instinctively feel it is true. My belief in causality is a matter of hunch. I live by that, while being aware that it may not be true.

To continue, even if causality exists, there is no foolproof way of discovering the actual laws of causality. In economics – and, with a longer history, also in epidemiology – by using randomized control trials, we try to infer causality. However, as some practitioners of randomized trials know (though, sadly, many do not) it is impossible to establish causality with certainty. Even for our own body, mind and behaviour, there is no way of being sure of how causality works. The best statement of this that I have heard comes not from a philosopher or a mathematician or an economist but a poet. Philip Larkin put it beautifully in a 1982 interview with the *Paris Review*, while talking about the challenges he faced in school because of his stammer and his shyness: 'I often wonder if I was shy because I stammer, or *vice-versa*.'

Once you accept causality, as most reasonable people do, determinism follows as a logical consequence. If two individuals make different choices in a similar situation – say, one commits murder and the other does not – the cause of this can be traced beyond the agency of the two individuals. Since human life is finite, we can trace the cause of their different behaviours to factors beyond them – either some difference in each of their environments (they have no control over that) or differences in their genetic make-up (they have no control over that either). Therefore, our behaviour, down to its minutiae, is determined by forces beyond our control.

One question arises from this that troubled ancient philosophers: if we are fully determined, can we have free will? As the philosopher Rachana Kamtekar asks, '[Could] Aristotle endorse the Kantian idea that even if theoretical reasoning leads us to conclude that we are entirely determined, nevertheless from a practical standpoint we must think of ourselves as free?'[10] The qualifier 'from a practical standpoint' suggests that there is an inconsistency between determinism and free will, which we, to be practical, must overlook.

It is, however, arguable that there is no real conflict between determinism and free will.[11] Suppose you choose not to study for an exam and then you fail. Determinism would say, and I would agree, that it is predetermined that you will fail. The mistake is to argue that if you are predetermined to fail an exam, you will fail the exam no matter what you do. For those who subscribe to this view, determinism is the same as fate. Your destiny is written down at birth and you cannot change this by your actions. There is a logical slip in reaching this conclusion. In reality, it is possible for both propositions (a) and (b) to be valid, where

(a): If you study hard you will pass the exam.
(b): It is predetermined that you will fail the exam.

In other words, it is legitimate to say you have the freedom to study hard and, if you do so, you will pass the exam. Determinism does not deny this. It simply says that it is predetermined that you will choose not to study hard. If an observer knew all the laws of nature, they would know which choice you would make. But that person's ability to predict your choice does not mean you did not have one. You had a choice, and the choice could have made a difference to the outcome. This is important because it means that you have moral responsibility for some of the things that happen in the world.

Let's go back to the description of a game from Chapter 2. In a game, we describe each person's set of available strategies or actions from which the person has to choose. This is referred to as the person's feasible set, or set of strategies. In the above discussion of my college days in Delhi, quite apart from my doubts about what a Communist revolution would achieve, I could not see that there was any action in my feasible set, or set of strategies, that could bring about a revolution.

These pure definitions segue into important moral dilemmas. Imagine a swimmer is drowning in the sea while you are sunbathing on the beach. You hear the call for help and it is entirely within your capacity to jump into the sea and save the swimmer. You are a good swimmer; you may bruise and hurt yourself but you will not drown. If, in this situation, you choose not to jump into the sea, and she drowns, you have a moral responsibility for her death. There is scope for some argument here about the use of the word 'moral', given that by determinism we know that this person would choose not to jump into the sea because of causes that go beyond the person. I need not go into that here. If a person has an action in his feasible set that can bring about a morally good outcome and the person chooses not to take that action, we shall follow the convention of saying that the person has moral responsibility for the bad outcome.

Even with this definition of moral responsibility, there are huge swathes of life involving large groups where evil can happen without us being able to apportion moral responsibility to anybody. As we shall explore, this raises difficult questions about contemporary politics, conflict and how to create a better world. Most people still find it difficult not to attribute agency and responsibility for things that happen in life.

Suppose we have a nation in anarchy, which leads to famine, human suffering and deaths. Suppose also that if the government had acted differently, this would not have happened. Our tendency is to hold the leaders in government morally responsible for the suffering. Is that right? Suppose no leader does anything because unless all leaders act together, nothing can be done. In such a situation, there is no compelling reason to think of the leaders as morally responsible for the suffering of the citizenry.

You may try to counter this by asking, *Why didn't any leader call on all the leaders to act together?* If such a call could lead them to act in unison, then of course this leader would have moral responsibility, because she does have an available action – in this case, a speech act – by which she can stop the nation's suffering. But I was considering a situation in which 'no leader does anything because unless all leaders act together, nothing can be done'. If this is really true, then no matter how significant the human suffering, we cannot say the leaders bear moral responsibility in any obvious sense.

I should point out here that this does not mean that you, the observer, should not *say* they have moral responsibility. If your saying so makes all the leaders act and this stops the suffering, then, even though the leaders do not have moral responsibility for the suffering, you have a moral responsibility to say that the leaders have a moral responsibility.[12]

Determinism has philosophical implications for how we lead our lives, how we think and how we apportion moral responsibility among human beings for what happens in the world. But before getting into that, I want to draw attention to the fact that there is a certain poetic beauty to observing the world with a determinist's eye.

Consider one implication of determinism. If, at some point in the future – say, after a zillion years – the universe happens to be exactly like it is today, this immediately implies that, after every zillion years, the universe will be exactly like today. The logic is easy to see. If from today the world goes on a path so that after a zillion years it looks exactly like today, then one day after one zillion years it will look exactly like one day after today, since the causal priors are identical at these two points of time. This means the process now becomes inexorable, leading us in a loop to the same universe after the same stretch of time. That this is entirely possible is a beautiful thought.

Just as Socrates, on that spring morning in 399 BCE, walked through the bylanes of the agora, stopping to talk with ordinary Athenians, unruffled by the fact that he was walking to appear in the People's Court, and just as five hundred men, who were members of the jury, also converged on the same court in the agora, Socrates would again, a zillion years later, on an identical spring morning, walk through the bylanes of the agora, stopping to talk with ordinary Athenians, and once again the five hundred members of the jury would be traversing the same path to the same court with the same thoughts on their minds. And they would do the same again every zillion years.

Why just Socrates and the jury? I will be back again a zillion years later, asserting that I may be back after every zillion years. And you will be reading a book and wondering if the author is right and you will be reading the same book after a zillion years.

It is entirely possible that the universe is an endless loop. The universe may eventually collapse, leading to a Big Bang, leading us back to this moment. Perhaps there is nothing more to life than that. The same you, me and all of us will be back again and again, like a refrain in a poem – lines or words that keep coming back, creating a hypnotic rhythm. If this thought gives you a certain poetic calm, it will give you the same poetic calm every zillion years. There is no more meaning or significance to life. If this infinite loop were true, and it is entirely possible that it is, it has no implication for us and how we live. It is just a matter to brood over – the inexorable wheel of the universe.

It is quite remarkable that some early Greek Stoics, most

notably Chrysippus of Soli, argued similarly: that the universe's trajectory over time may be endless, with no beginning, no end and no meaning. Whether they reached this conclusion deductively from the axiom of causality and determinism, as seems likely, or by some other route, there is no way to be completely sure. The audacity of the ancient thinkers is something to marvel over. It is reassuring that the line taken later in this book – namely, that of treating determinism as compatible with choice and moral responsibility – was a part of this early philosophy.[13]

Philosophy is typically written in analytic prose, but we should be remiss not to recognize that it can also be captured as evocation, as in Rabindranath Tagore's poetry, Bob Dylan's lyrics or that magical song by Pete Seeger, 'Little Boxes', which captures the same idea of Chrysippus – of repetition and the strange beauty of monotony:

Little boxes on the hillside
Little boxes made of ticky-tacky
Little boxes
Little boxes
Little boxes all the same.[14]

The value of regret

Let me return now to the functional part of determinism, its contribution to our way of life. As noted earlier, determinism helps us understand that what others do, they do because of causes that transcend them. We never get angry at tigers even though tigers bite, because we realize that that is the way tigers are. This intellectual understanding and resulting equanimity, in turn, helps us deal with tigers rationally and more effectively, whether in running away or in striking back. We are much worse at handling human beings. Our anger and hatred often overcome our judgement and our ability to make reasoned decisions. Determinism tells us that we should show the same understanding towards humans; for the way they behave is, ultimately, caused by factors beyond them.

Just to clarify, determinism does not mean you do not punish but that you do not suffer from the emotion of anger. Further, the punishment should not be for revenge. It is to make the one being punished and the observers of the punishment behave better in the future. Punishment must not be an act of vengeance but an act to create a better world by changing behaviour in the future.[15]

Determinism implies that only *you* can change the course of the universe. There is an echo of this philosophy in the practical writing of the celebrated human-rights activist Maria Ressa (2022, p. 56), when she says, 'Learning to be honest begins with your own truths: self-assessment, self-awareness, your empathy for others. *The only thing you can control in the world is you*.' (My italics.)

Each person has to understand that, apart from what he or she can do (the choices available in his or her feasible set of actions), everything else in the world is given. From your point of view, either you are responsible for what is happening in the world or there is nothing to be done. There is no third option. The thought that if another person did something that would lead to a better world is of value only if there is something that *you* can do to make the other person do that thing.

If your speech, your protest, your vote, a change in your career and so on cannot help society be better, then your society not being better is a fait accompli, like gravity. You may hate gravity for pulling you down, but there is nothing you can do about it. That being so, hating gravity is not a good idea. It is getting yourself riled up for no purpose.

There is a similar lesson here about regret and remorse. In varied degrees, most of us have these emotions and some suffer greatly from guilt feelings. Why did I hurt this person? Why did I do what is obviously wrong in the past? These emotions are there in us for a reason. They help us to become better people over time. Evolution likely endowed human beings with these emotions as it made them more capable of improvement. However, even if remorse and regret play a role, there is really no reason to have remorse and regret. This is because your own past actions are part of the structure of the world. There is nothing you can do to change them. Your past action is like gravity. You may like it or hate it; you cannot change it.

From this follows a simple lesson that echoes the earlier insight about anger. We should try to let go of the emotions of remorse, pain and regret, but, while doing so, we need to retain the good that comes from those emotions. The way to handle this is to tuck the information about one's past behaviour into the intellectual part of the brain.[16] You must not deny that a wrong you did in the past was a wrong, but keep this stored in your intellect, to guide better behaviour in the future.

So we return to where we began (perhaps it was predetermined). Philosophy, nowadays, is treated as a scholarly discipline, something that is to be learned, something for which one sets aside time, as one does with various courses as a student. Philosophy is found in books, but it is also a way of life. And it is the latter that is most valuable. It teaches us to ponder and reason, even as we go about our days. It is that capacity that enables us to lead the good life. The most striking example comes from Socrates' life. He did not have an academy. He taught, learned and practised philosophy as he went about his daily routine, walking the bylanes in and around the agora in Athens. Some of his most important discussions happened in the home of a cobbler, Simon the Shoemaker. As Bettany Hughes (2010, p. 23) puts it, 'A cobbler's workshop-home would have seemed the most appropriate of places for unconventional Socrates to analyse the meaning and point of our everyday lives.'

Taking my cue from the above discussion, I shall assume that each person is responsible for choosing whatever they can select from the feasible set of choices available to them. With this assumption in place, we are now ready for one of the most contentious topics in philosophy: the conception of the good life and group moral responsibility, the subject matter of the next chapters.

5

Greta's Dilemma

The invisible hand of malevolence

What gets little attention in mainstream economics but is an inherent part of the human mind and provides some of the critical foundations of economic growth is the moral intention. By morality I am referring to an innate sense of fair play and justice. If there were a God who rewarded good behaviour and punished bad, there would be no need for morality. Self-interest would do the job. In reality, we need morality.

Fortunately, we all have a moral compass in our heads, which includes a sense of empathy. The extent varies and our predominant concern is usually with our own well-being, but virtually all human beings put some weight on the well-being of others, and have a yearning for fairness. We want others to be fair towards us and want to reciprocate that. At times, this takes the form of lip service. We are not fair but create the appearance of fairness. However, even this urge to camouflage our selfishness is a sign that we value fairness.

Even though some animals also exhibit some of this trait, it is a defining feature of human beings. Human progress and economic growth, in large measure, were made possible by the possession of a moral compass. Rats have self-interest as much as human beings but not enough empathy and moral intention to form a meaningful community of rats. The reason why the world of rats does not have the invisible hand of the market enabling trade and exchange, and the reason why economic growth is meaningless when applied to rats, is not the lack of self-interest among rats but the lack of an adequate hard-wiring of norms of fairness and justice, so vital for the invisible hand to function and for progress to be possible. These are assumptions that some neoclassical economists do not recognize, but are in the woodwork of economics, as discussed earlier.[1]

The traditional game theorist may respond by pointing out that, though game theory does not talk enough about our moral yearning explicitly, this is allowed by the standard paradigm of game theory. This is because the payoff the person gets is a 'primitive' in game theory. That is, it is not for the game theorist to decide what gives individuals their payoffs. It is entirely possible that a part of the payoff comes from our eating apples and owning gold, but a part also comes from seeing others eat apples and have some wealth.

This is a valid response, but the counter to this is to argue that by treating the payoff function as a given, we lose out in important ways. There are insights to be gained by parsing out our preferences into the selfish and non-selfish parts. This allows us to see how we may end up building an unfair society that none of us may individually want. A sad fact of the world is that there are many evils for which no one is responsible. Even though collectively we could stop the evil if we wanted to, no *one* has the power to do so. This is the chilling world Franz Kafka described in his fiction, most notably in *The Trial*, and in *The Castle*.

In *The Trial*, one morning, like every other morning, soon after eight o'clock, Josef K. waits in his apartment expecting his landlady's cook to bring him his breakfast. Instead, he is arrested. It is not clear what the charge is or who, for that matter, these policemen are, who came in unannounced with a warrant. 'What sort of people were these? What were they talking about? What office did they belong to? K. was living in a free country, after all, everywhere was at peace, all laws were decent and were upheld, who was it who dared accost him in his home?'

These questions never get answered. All individuals go about doing their daily jobs, creating *in their collectivity* a system of oppression, but it is never clear who the oppressor is or whether, for that matter, there is anybody directing this society. It is the auto-totality of Kafka's oppressive system that is frightening.

There are echoes of this in the real world of post-totalitarian dictatorship that Kafka's countryman Václav Havel describes in his samizdat pamphlet, aptly titled *The Power of the Powerless*. 'The manager of a fruit and vegetable shop places in his window, among the onions and carrots', a poster that declares loyalty to the oppressive, totalitarian regime of the country. 'Why does he do it? What is it that he is trying to communicate to the world?' asks Havel.[2]

Is he really loyal to the regime? No, answers Havel. He puts up the poster because it was delivered to him from 'the enterprise headquarters along with the onions and carrots'. If he did not put this up, he would be viewed as 'disloyal' and he would be harassed. He puts up the poster purely for that reason, not because he *is* loyal.

We have seen peer pressure like this work in many situations, from the school playground and office politics to the practice of casteism in Indian villages.[3] But what makes Havel's analysis so deep is that he does not leave it at this. He asks the next obvious question. Why do those who would harass the greengrocer for being disloyal do so? The answer is the same. If they did not harass someone signalling disloyalty, that would be a signal of their disloyalty and they in turn would be boycotted and ostracized. In the end, the whole system is locked in a 'blind automatism'. All individuals, from the greengrocer to the party bosses, are locked in an equilibrium. All individuals are simultaneously both 'victims and pillars of the post-totalitarian system'.

Even though there are societies where we can point a finger at the evil perpetrated by the leader, it is possible to have systems – and this may even be the norm – where the leader is also a victim of the system. There is nothing that they can do without disproportionate harm to themselves (and protecting oneself from harm beyond a certain level is understandable behaviour even by a moral person) to help society break out of this equilibrium. Havel's booklet is a political manifesto but also an exercise in contemplation about society. As if to confirm his ideas, soon after the publication of this booklet Havel was imprisoned in Czechoslovakia, a nation that was a mirror image of what he had described.

Clearly, the world that Kafka and Havel describe is one driven by an invisible hand. The oppression cannot be attributed to any single person. It is the outcome of collective behaviour. It is *the* invisible hand of malevolence. What is fascinating is that, at a fundamental level, it is the same idea as that of the invisible hand described by Adam Smith in 1776. Smith's invisible hand of the market delivers order and efficiency. Smith showed that to have an optimal market it is not necessary to have a prime minister, president or God. The order can be a result of a blind automatism, of the same kind as that described by Kafka and Havel.

The discovery of the invisible hand by Adam Smith was a major breakthrough, but the works of Kafka and Havel are reminders that the invisible hand does not *have to* work for the benefit of individuals. It can be a force of oppression and suffering. We need to be vigilant and understand these invisible forces that hold society together and can turn from being a force of good to one of evil. This disconnect between individuals and groups is what Nietzsche was drawing attention to when, in his *Beyond Good and Evil*, he wrote, in his characteristically flamboyant style, 'In individuals, insanity is rare; but in groups, parties, nations and epochs, it is the rule.' The next section illustrates how we can give formal content to Nietzsche's conjecture.

Greta's Dilemma and group moral responsibility

Game theory has vastly improved our understanding of group behaviour and dynamics. We are today better able to address many of our collective failures. When we see a group of people overgrazing the commons or damaging the immediate environment, we do not say that they should learn to act in their self-interest because what they are currently doing is hurting their own interest. Thanks to games like the Prisoner's Dilemma, we realize that it is, in fact, their self-interest that is causing the problem. We need conventions and collective agreements to restrain our immediate interest and thereby serve our collective, long-term interest, and we also need to understand how conventions and collective agreements can be held in place. Game theory has played an important role in the design of conventions pertaining to climate change, trade policy, labour regulation and even monetary policy, which helps restrain narrow self-interest for our larger collective interest.

However, when it comes to moral pursuits, our understanding remains rudimentary. This understanding is crucial to creating a better world and realizing many of our moral intentions. We human beings have innate moral intentions, but we must still understand how these intentions may fail to be realized and may even backfire.

Our propensity to attribute agency to groups often leads to a strained search for their 'hidden agenda'. More often than not, groups do not have hidden agendas. Individuals comprising the group may have such agendas but often the behaviour that emerges from the group fails to reflect the agendas of individuals.[4] There are fewer conspiracies in the world than people imagine.

We see group moral attributions almost every day in newspapers and magazines, on television and, of course, on Twitter and Facebook. Thus, in the US, we hear references to 'the damage being done to our country by the immoral Democratic party'.[5] Similarly, we read about how the 'Republican elite's immorality goes well beyond Donald Trump . . . The immorality and disdain of today's Republican elites shine through in the policies that they embrace.'[6] Observations like these, while they may be accurate in some cases (I must confess to a hunch that one of the two above statements is accurate), do deserve scrutiny.

Group moral attributions are surprisingly common. We talk about the lack of morals of big corporations; we use collective moral epithets to describe the Houthis and Tutsis, the North Korean leadership, FARC rebels and protest groups, as in the Arab Spring or Hong Kong's Umbrella Revolution. After the Pulwama attack in India, on 14 February 2019, when a terrorist group operating out of Pakistan attacked an Indian convoy, there was an emotional discussion on who bore moral responsibility for this act – the one who carried out the attack, Pakistan's leaders or all Pakistanis.

Group moral responsibility is a philosophically contentious topic.[7] In many cases, group moral attributions may be meaningless and, in whipping up anger towards groups – nationalities, races and religious groups – they may even be dangerous. Admittedly, at times it may be worthwhile to say that individuals comprising a group are morally responsible even when they are not,[8] since saying this can inflict guilt and incentivize people to behave better. In short, there may be consequentialist reasons to *say* so even when it is not right to *think* so.

Human beings have a proclivity to attribute agency whenever they see anything happening. Whether it be a flood, drought, war or conflict, we like to think of someone responsible for it. It is this propensity that has led many people to believe in gods. It is the same propensity that makes us hold groups collectively responsible for what happens in the world, without pausing to ask if such attribution has any meaning. The *Economist* magazine, 16 April 2022, illustrates this with the cover story, 'What China is getting wrong'. It is difficult to think of a larger group to which to attribute agency. At one level, this is simply a manner of speaking, but it is one that influences our thinking and our behaviour, leading us to jump from observed behaviour to attribution of group moral responsibility.

As we have already seen in earlier chapters, it is valid to question whether individuals are responsible for their choices given that the choices they make are predetermined, by their own genetic make-up and environmental factors. There is plenty of debate about whether a person choosing from among a set of alternatives can be held responsible for his or her choice.[9] I shall stay away from that debate here and assume that each person is indeed responsible for the choices they make from within the set of actions available to them. Even without shying away from this attribution of responsibility, new problems arise in strategic environments involving groups of people.

Consider a new version of the Stag Hunt game discussed in Chapter 2. Suppose that there is a large number of players – let us assume 100 of them. Each person has to choose between strategies A and B. If all 100 choose A, they get $101 each. If all choose B, they get $100 each.

If not all players make the same choice, the payment is made as follows. If n of them ($n < 100$) choose the same action, they get n cents each. Thus, if 60 players choose A and 40 choose B, all the ones choosing A get 60 cents each and all choosing B get 40 cents each.

It is easy to see that this game has only two Nash equilibria: every player choosing A or every player choosing B. It seems reasonable to suppose that, if this game is played repeatedly, the players will settle into all playing A or all playing B. Once they settle into one of these outcomes, no one will have an incentive to unilaterally change their behaviour. Of course, if you are in the equilibrium of playing B, you may regret that you are not in the 'A equilibrium', where you would get an additional dollar, but there is little you can do individually. For any other outcome – that is, apart from all choosing A or all choosing B – you can do better by an individual deviation and so none of those constitutes an equilibrium.

Now suppose that, apart from the players, this society has 'bystanders'. These are individuals who are affected by what the players do but can do nothing to affect the payoffs of the players or even their own payoffs (the latter is an assumption made to keep the argument simple). A typical bystander in today's world is the future generation. What we, the 'players', choose today can affect the well-being of the future generation, but nothing the future generation does can affect us. This is also reasonably true of the poor and marginalized sections of a contemporaneous society. For instance, the actions and the choices of individual landlords in a feudal society have a large effect on the lives of the serfs, but the actions of individual serfs have a negligible effect on the well-being of the landlords. Throughout this section, in talking about games, I will think of a 'society' as consisting of two distinct kinds of agents: the players and the bystanders.

Let us suppose that in addition to the 100 players, there are 1,000 bystanders. They are much poorer than the players. Their payoffs, or earnings, are determined as follows. If all players choose B, each bystander gets $5. If they all choose A, the bystanders get $1. For all other outcomes, they get nothing. Let us suppose an income of $1 is a life of misery, whereas $5, while still poor, is tolerable.

Suppose this society is settled in the 'A equilibrium', that is, all players play A. For outside observers, this society will seem like a mean and morally irresponsible country. They will talk of another, otherwise identical society they have seen, where everybody plays B, and, in comparison, will talk of the players in the A equilibrium as morally depraved. There will be columnists writing about the damage being done to the poor bystanders by the immoral citizen-players of this nation, how just to earn one extra dollar these rich people are willing to push poor bystanders into extreme poverty and to living on one dollar.

However, we know in these cases the group moral attributions are wrong. We can have two similar societies, indeed two societies with exactly the same payoffs, in which each player tries to maximize their own welfare. One of them could settle into the A equilibrium and the other in the B equilibrium. The outcome has nothing to do with their moral intention.[10] So, while it is indeed true that one outcome is good and one bad, it is not always the case that we can blame the players of one society (or praise the others).

The problem can be more acute than this. There are contexts in which a society's collective behaviour could become morally worse by virtue of one or even all the players becoming morally better.[11] This is a big challenge for policymaking and collective action. For individuals to be moral does not mean that the outcome will be more moral. Even if people share Greta Thunberg's moral intention to act so as to help our future generations lead better lives, we may end up hurting the future generations. This is what may be referred to as 'Greta's Dilemma'. Just the *intention* is not enough.

Before exploring this formally, I want to stress that this is much more than an academic matter. History is replete with examples of good intentions leading people to failure. The best example may be from the utopian socialists, like Henri de Saint-Simon, Charles Fourier, Robert Owen and several others, who tried to build utopian communities in Europe and the United States in the eighteenth and nineteenth centuries. Their sense of disquiet about the unfairness of the world around them, as European society was thrown into chaos by the Industrial Revolution, cannot be faulted. Nor can we dismiss their aspiration for a better society. There was something ethically magnificent in their vision. Yet, they failed, often leaving wounds deeper than before they began. Saint-Simon, for instance, proposed, prophetically, the idea of a union of European nations as early as 1814, but his detailed plan was unpractical, and it came to naught. Three years later, he began writing about his vision of a socialist utopia. But his obsession with building canals across a utopian world proved to be a distraction that led to failure. Disappointed, and impoverished by having nobly put down his money to try to build his utopia, he, finally, tried to shoot himself on 9 March 1823. But, once again, he failed. He just lost an eye.

History has other, more momentous examples of utopian experiments that failed. In 1958, Mao Zedong began his Great Leap Forward programme in China, whereby individual property rights were taken away from farmers and they were organized into large cooperatives. Their individual returns (or payoffs) were no longer connected to their effort. This utopian construction, underestimating the role of incentives, backfired, unleashing on China the biggest recorded famine in human history, with around thirty million dead.

Moral intention on its own is not enough: it needs to be paired with hard-headed science and the mindset of creative research. The Greta's Dilemma game that I am about to describe is an allegory to illustrate the importance of this.[12]

Consider the interaction between two individuals, Johnny Won and Jaya Tu, whose names, luckily, can be shortened without much phonetic distortion to Players 1 and 2. In this game, Player 1 chooses between actions A and B, and Player 2 chooses between a separate set of actions, C and D.

To give a story to the symbols, think of Johnny Won (1) as choosing between organic 'agriculture' (A) and the more damaging business of operating 'brick kilns' (B). Jaya Tu (2) has to choose between 'coal mining' (C), which can do a lot of damage to the environment, and D for the greener (albeit not carbon neutral) activity of 'dairy farming'. The significance of these will be obvious when I consider the fallout of these choices for the bystanders – in this case, the future generation.

The payoffs are as follows. If Player 1 chooses organic agriculture, he will always get $100; if he chooses brick making, he will always get $101. For 1, B is the dominant strategy:

no matter what 2 does, from 1's point of view brick kilns is a better business proposition than agriculture. However, Player 2's payoffs depend on what 1 does. If 1 goes for agriculture, it is known that this will be predominantly soya production, which is a kind of substitute for dairy farming. So it is more profitable for 2 to go for coal mining (a payoff of $101 instead of $100). But if 1 chooses the brick business, the demand for dairy will be high and 2 will prefer dairy farming (again, a payoff of $101 over $100). Don't worry if this is hard to hold in your head all at once, as we will summarize it in the table – referred to in game theory as a 'payoff matrix' – below.

Given individuals' wish to maximize their own earnings, it is easy to see that in this game the only equilibrium is (B, D), where they earn (101, 101). Both players stand to receive the maximum profit. If 1 deviates to A, he earns $100, and if 2 deviates to C, she earns $100. So neither will deviate.

Now suppose this society has a future generation that is poor (thanks to the damage to the environment done by the present generation). The future generation's well-being is entirely dependent on what the current generation does. If 1 and 2 choose the environmentally friendly actions A and D, the future generation gets $8. If they choose the environmentally worst option (B, C), the future generation gets nothing. If they choose the environmentally more mixed options (B, D) or (A, C), the future generation gets payoffs of $4 or $2.

Here are the payoffs of the players represented in a payoff matrix (on the left), and also the bystanders' (the future generation's) payoffs (on the right).

The Basic Game

	C	D
A	100, 101	100, 100
B	101, 100	101, 101

Future Generation's Earnings

	C	D
A	2	8
B	0	4

Let's assess the likely outcomes. In standard game theory and neoclassical economics, players make their decisions with their own interest in mind and the future generation's payoff is of no consideration to them. Note that this leads to the equilibrium (B, D), where the two players get $101 each. This is the only outcome from which no individual can unilaterally make a change and benefit by this (which is the central idea of Nash equilibrium in a game). The players get $101 each, and the bystander gets a miserable $4.

To most observers, the behaviour of the group of players will appear unethical. The current generation is well off, with each person earning at least $100. Moreover, how they choose can make only a $1 difference to their income. The future generation's condition is likely to be much more precarious, with much lower payoffs. Note that if the players moved from (B, D) to (A, D), the future generation, or the bystanders, would get $8 instead of $4. It appears unconscionable not to make this small sacrifice.

Now suppose Player 1 has a chance meeting with Greta Thunberg and learns about the importance of the future generation, who does not have a voice in our current deliberations. Player 1 learns that a minimal requirement of morality is to be mindful of how one's actions affect poor bystanders. He takes Greta's advice to heart, as indeed all of us should, and begins to take the welfare of the future generation into consideration in making his decisions. For simplicity, assume Player 1 now gives the same weight to the future generation's earnings as he does to his own.

This clearly causes the game to change. There is now a selfish player (Player 2), whose payoffs are as before, and a moral one (Player 1), whose payoffs are his own earnings *plus* the earnings of the future generation. This is represented in the payoff matrix below. Player 1 now tries to maximize not his earnings but the aggregate of his earnings and those of the bystander. I call this new game that emerges after one player turns moral 'Greta's Dilemma'.

Greta's Dilemma

	C	D
A	102, 101	108, 100
B	101, 100	105, 101

What is the outcome of Greta's Dilemma? It is easy to see it has only one equilibrium: (A, C). The old outcome, (B, D), is no longer an equilibrium, since if Player 2 chooses D, Player 1 will choose A in order to help the future generation. And knowing that Player 1 would choose A, Player 2 would choose C. Check that no other pair of choices is stable in this sense. We invariably end up at (A, C), thanks to Player 1's conversion from an amoral agent to a moral person.[13]

At first it may seem that the world of the game has improved, since we have progressed from a world where every player is selfish to a world where only half the players are selfish. But what happens to the future generation in practice? Before Player 1 met Greta, the future generation earned $4. After 1 comes under Greta's influence and becomes an environmentally conscious person, determined to help the future generation, the future generation's well-being ends up dropping to an even lower payoff of $2.

Outside observers, seeing the atrocious behaviour of these two rich people pushing the precarious future generation to receive a measly $2, will find it difficult to understand that this is happening *because* one of them has become moral, with a commitment to help the future generation. Clearly, the moral intention on the part of an individual, in a strategic environment, does not necessarily translate into a moral outcome.

To understand Greta's Dilemma, let us agree that an outcome in which anybody gets a payoff of less than $3 is a 'bad outcome'. Thus (A, C) and (B, C) are bad. It is easy to see that if a bad outcome occurs, Player 1 has no responsibility for this, since the outcome would be bad no matter what he chose. But Player 2 does have moral responsibility. If she opts for D, the bad outcome can be averted for certain.[14] The problem arises from trying to reconcile this with the fact that it is Player 1's meeting with Greta and *becoming moral* that causes the bad outcome.

One possible way of countering this is to argue that Player 1 can surely see that it is his being moral that causes the harm and so it is better to *act* amorally. Then, the game will end up at (B, D), where the bystander does better by getting $4. But this is akin to arguing that, confronted with the Prisoner's Dilemma, players will have the sense to behave as though they are not selfish and achieve the good outcome. Such an argument is dismissed by theorists on the grounds that it always pays to deviate after persuading the other player that you are someone else.

The analogue of this in Greta's Dilemma is for Player 1 to pretend to be amoral, drive the game to (B, D), and then make a last-moment switch to A, to achieve (A, D) and help the bystander get $8. To this the game theorist's rebuttal will be: a rational Player 2 will expect this and so play C in anticipation, and we are back to (A, C). The advantage of such abstract analysis is that it is a tool that can be adapted to many contexts. The story could be about soldiers in an army, who are initially unmindful of the loss of lives on the side of the enemy, before some develop concerns about committing violence without just cause. The story could be about the wealthy making some small sacrifice for the poor.

There is an interesting difference between the selfish individual of traditional game theory and the moral individual we are discussing here, in terms of commitment to certain kinds of behaviour. In game theory, we treat self-interest (that is, the urge to maximize one's own payoff) as innate. But it is arguable that being moral entails an element of volition that goes beyond this. So it is possible to argue that individuals can *choose* to be moral (or not), after examining the context of the game. If they are sufficiently far-sighted, they may decide, in certain contexts, not to *behave* morally, if that leads to a moral outcome.[15]

Greta's Dilemma also takes us beyond moral philosophy to connectionism in the cognitive sciences, which recognizes how one human mind consists of large networks of neurons, each neuron oblivious of its role. There is now an effort to bring this to the social sciences and recognize that there may be large networks of humans behaving collectively in ways that each person is unaware of.[16] In studying such collective organisms it will, in many situations, be futile to morally evaluate individuals because they may not have volition in the conventional sense.

I would argue that, for now, hope has to lie in the kind of suggestion put forward by the philosopher Hilary Putnam, drawing on the work of the French-Lithuanian philosopher Emmanuel Levinas: 'For Levinas, the irreducible foundation of ethics is *my* immediate recognition, when confronted with a suffering fellow human being, that *I* have an obligation to do something. [Even if I cannot actually help,] not to feel the obligation to help the sufferer at all, not to recognize that if I can, I must help . . . is not to be ethical.'[17] Note that he is not violating the dictum *ought implies can*. He is not asserting that you ought to help someone whom you cannot help, but you ought to *feel the obligation* to help.

Even if a good outcome is beyond our reach now, we must nurture and keep alive what may be called the 'moral intention', which is the intention to achieve the good eventually. It is our moral intention that makes us want to step beyond the game under consideration and think of how we may alter our behaviour. The moral intention *can* backfire and make matters worse, as we just saw with Greta's Dilemma, but, nevertheless, it can inspire us to look beyond the game and think of novel solutions, such as imposing taxes or fines on selfish players. I shall comment on this later in the book.

There is no easy getting away from the core message of Greta's Dilemma. When we see a group behaving badly, we must not assume that the outcome reflects what the individuals in the group desire. Most people do not allow the thought that a group of leaders may not individually want to do what the group often ends up doing. They may be caught in the same trap that Václav Havel imagined in his post-totalitarian state. This is indeed a dilemma of which Greta Thunberg, with her good moral intention, has to be aware. Individuals who are concerned about future generations not only may not help those generations, but may actually end up hurting them – and not because they failed to act morally, but because their moral choice is part of a bigger game that is actually played out.

The Samaritan's Curse

It is reasonable to ask if the counter-intuitive result illustrated by Greta's Dilemma is an outcome of the fact that only one person becomes moral. What if all players became moral? Could the moral intention still backfire with an immoral outcome? In Greta's Dilemma, as is easy to check, the counter-intuitive result could not occur if both players became moral. However, that consolation is not always the case in the big, wide world of ours.

With a little effort, we can find examples where every player turning moral will make the outcome more immoral. Consider a society with two players, 1 and 2, and with each player having three strategies or actions, A, B and C, to choose from. It is a harmless simplification to use the same letters to describe the strategies of the two players. Hence, the society has nine possible outcomes. Player 1 chooses between the rows and Player 2 chooses between the columns. The payoffs they earn are shown in the table, or payoff matrix, labelled 'The Basic Game', below.

The Basic Game

	A	B	C
A	102, 102	80, 120	108, 108
B	120, 80	104, 104	80, 102
C	108, 108	102, 80	106, 106

Bystander's Earnings Matrix

	A	B	C
A	20	4	0
B	4	6	10
C	0	10	4

It is easy to see that the only outcome where this society will end up is (B, B); that is, an outcome where Player 1 chooses her action B and Player 2 chooses his action B. This is a Nash equilibrium, since no player has an interest in deviating unilaterally. Let us check out any other outcome. Consider (A, A). Here, both players earn $102. However, each player can do better by deviating unilaterally. If, for instance, Player 1 chooses B instead of A, she will earn $120. The reader should be able to verify this is true for all outcomes except (B, B).

As in standard models of games, neither player pays any attention to the fallout of their behaviour on bystanders. Let the payoffs earned by the bystanders be as shown by the above matrix labelled 'Bystander's Earnings Matrix'. Since the equilibrium is (B, B), the bystanders get a payoff of $6 in equilibrium.

Now a Good Samaritan comes to town, is dismayed by the moral degeneracy of the people and teaches players basic morals.[18] Like Greta in the above section, the Samaritan tells them: Your choice of (B, B) gives you ($104, $104), but do you not see that this leaves the bystander with a miserable payoff of $6? If you, the super-rich, opted for (A, A), you would lose only $2 but the bystander would get $20. Surely you should be prepared to sacrifice $2 if it means an additional $14 for the poor bystander. Ignore the other player, if you wish, since she is super-rich like you, but be mindful of what happens to the poor when you choose.

Suppose now both players become moral creatures. Each values the earnings of the bystander (anyone who has an income of less than, say, $25) as much as his or her own payoff. So if the outcome is (A, B), Player 1 gets a consolidated payoff of $84, which consists of $80 of her own and $4 for the poor bystander, and Player 2 gets a consolidated payoff of $124. By writing in these consolidated payoffs for both players, we get a new game, the Samaritan's Curse. Exploring the possible outcomes shown in the payoff matrix below, you can see that the equilibrium of this new game is (C, C).

It is easy to see that (B, B) can no longer be an equilibrium. Player 2, for instance, now a moral creature, will deviate to C if Player 1 chooses B. The players' turning into Good Samaritans worsens the bystander's condition. The bystander was earlier getting $6, and now gets $4. Individually moral behaviour – even from every player – makes the group behave immorally.

The Samaritan's Curse

	A	B	C
A	122, 122	84, 124	108, 108
B	124, 84	110, 110	90, 112
C	108, 108	112, 90	110, 110

It is now clear that neither one person turning moral nor the whole group turning moral assures an escape from this paradoxical result.[19] These two games should warn us that when we look at a group's immoral behaviour and implicitly think of the individuals constituting the group as immoral, we may be utterly wrong.

Some may object to the above analysis by arguing that, in both Greta's Dilemma and the Samaritan's Curse, the players who are being described as moral are not truly moral. If they were, they should analyse the consequences of their behaviour and see that their initially moral choice is doing harm. This should prompt them to change their behaviour to be nominally selfish in order to achieve an unselfish outcome. As we saw in the case of Greta's Dilemma, the problem with this is that one would always be tempted to cheat and revert to a previous strategy after having duped the other person. But, knowing this, the other person will not be duped. We end up at the only possible equilibrium.

What the above games do is to remind us that, when it comes to policymaking, we need to do for moral behaviour what we have done for rationality. A major contribution of game theory was to demonstrate that individually rational behaviour may not lead to rational outcomes for the group. This has led us to introduce laws, taxes and rewards so as to bring individual behaviour into alignment with our group interest. A host of tax policies for preserving the commons, and global agreements on climate change, have been made possible by this kind of thinking. The two games above prompt us to think in similar terms to bring individual morality into alignment with group morality.

It is easy now to see how popular discourse on moral responsibility is often flawed. During British rule in India, there were many Britishers who said that their aim was to help develop India. This is commonly treated as hypocrisy because there is evidence that the Indian economy was drained during colonial rule. In other words, behaviour by the British players left the bystanders, the Indians, exploited. Putting it in those words immediately alerts us that, by the logic of the above games, we cannot from the outcome of exploitation automatically conclude that the intention of all the British rulers was to exploit. However, the colonial machinery often created 'guilt shelters', which facilitated exploitation and extortion. This is the subject matter of the next chapter.

6

Collective Action

Corporations as guilt shelters

Collective action is a familiar topic for economists. The reason to get into this here is that the above abstract analysis opens up some novel and contemporary challenges for group behaviour and collective action in the real world. I want to give readers a glimpse of these challenges and draw them into the discourse, including some open questions about the foundations of game theory.

Games like Greta's Dilemma and the Samaritan's Curse have many interesting implications. In this section, I want to show how they help us understand a common phenomenon, namely, how large organizations, like nations and corporations, can act as 'guilt shelters' for the individuals that belong to these organizations, and that is one of the reasons they outcompete smaller entities.

Think of Voltaire's celebrated observation: 'It is forbidden to kill; therefore all murderers are punished unless they kill in large numbers and to the sound of trumpets.'[1] Killing as part of a large army takes away the individual guilt and shame associated with murder. If I did not kill, my fellow soldier would; so there is no reason for me to feel guilt and desist from killing.

A lot has been written about how large corporations have paid inadequate attention to broader social responsibilities, including towards climate change and the environment, thereby harming not just future generations but contemporaries who do not come under the umbrella of the corporation. A part of this is simply a matter of impunity. They have the ability to do things without fearing reprisal from consumers or workers, who typically do not have comparable heft. However, in addition to this, the large and complex structure of corporations allows them to create 'guilt shelters'; that is, protect the individuals who are part of the corporation from feeling responsible and guilty for the behaviour of the corporation and the final outcome.

One reason a small firm or a self-employed entity behaves well with customers and workers is guilt and a sense of injured morality that arises from not delivering where one has the capacity to deliver. There is now recognition of this in behavioural economics: we do not like to hurt and cheat. Beyond the self-interest of maximizing our own payoffs on which game theory is based, we also have a moral compass and sense of guilt that hold us back from certain socially irresponsible behaviours.

One reason why a customer dealing with small firms or an owner-operated firm is less likely to be treated poorly is that the responsibility for the poor treatment can be easily placed on an individual, and the individual will feel a sense of guilt, and therefore may exercise self-control. In a large corporation, it is often not clear who is responsible. Let's say that, as a customer of a large bank, we are treated unfairly or charged wrongly or sold a poor-quality product. When we phone to complain, it soon becomes clear that the person answering the call is just a small cog in a large organization, who had no role in the poor treatment you received. It is awkward for them to apologize for something they did not personally do. Equally, we find that we restrain ourselves, holding back our anger or irritation, on the grounds that this individual had no role in the poor behaviour of the corporation. If we were to spend time looking for who is responsible, on most occasions we would be disappointed. In the long supply chain that finally brings the product to you, there is often no *one* who can be held responsible, akin to what happens in Greta's Dilemma and the Samaritan's Curse.

This, in turn, allows and even prompts larger corporations to lower the level of treatment given to consumers and, for that matter, to workers. It allows them to damage the environment and cause a large negative fallout among wider society, because each individual in the corporation feels powerless under the circumstances. It is in the same way that armies cause collateral damage to people. Cities are often destroyed by armies in the course of war. Our natural sense of guilt and moral compass that would prevent us from such behaviour is sheltered under the cover of a large group being collectively responsible without being able to apportion guilt to us individually.

Of course, we can and do have laws to mitigate this effect. But the fact remains that large corporations with long and complex supply chains provide a guilt shelter to individuals, allowing them to behave in ways they would not on their own. These guilt shelters in turn enhance the profit earned by the corporations, enabling corporations to outcompete smaller firms and individual sellers.

It is not necessarily the case that corporations create these large and complex decision-making structures deliberately to create guilt shelters. It is possible that, by evolution, such corporations are more likely to survive and therefore gradually become the norm rather than the exception. However, we must not rule out the possibility that there are sufficiently talented corporate heads who realize the power of guilt shelters and are themselves morally brazen enough to deliberately create structures in which no individual employee of the corporation (other than them) has responsibility for the collective behaviour of the corporation.

Oppression and the Incarceration Game

Once we enter the domain of moral considerations in strategizing how we play games, a host of important topics open up, which prompt us to think of actions to prevent unwanted collective behaviours. Akin to the corporate head who designs a corporation so as to shield individuals from having moral responsibility, here is an illustration of how design and strategy can be used by some people to oppress the masses, thereby raising the question and forcing us to think of how we may prevent such behaviour. This is an interesting example because it illustrates the powerful real-life insights one can get by bringing in modern game-theoretic analysis to bear on abstract philosophical ideas. The rest of the section demonstrates this with O'Connor's (1948) Surprise Test paradox, which generated a lot of interest among philosophers and logicians.

The world has seen revolts and civil wars since ancient times. To understand revolutions, chaos and conflict is as fascinating as trying to understand the order that prevails in society, whether because of the invisible hand of the market or the authoritarian hand of the leviathan. However, in contrast to order and harmony, revolt is difficult to model. But we have made big strides in understanding anarchy and the state of nature, and there is hope for understanding revolt too.[2]

History is full of examples of revolution or civil war breaking out amidst seemingly peaceful times, and rebellion being quashed on the verge of success. Even within the last few decades we can find ready examples, from the Jasmine Revolution in Tunisia or the more widespread Arab Spring in the Middle East that began in 2010 and led to the collapse of several authoritarian leaders, to the 2020–21 protests against Lukashenko's brutal regime in Belarus or the protests in Iran sparked by the killing of Mahsa Amini in 2022. This may take the form of a gathering storm and then a sudden quietening down, with the streets abandoned by protestors because of the looming threat of reprisal, as happened in Myanmar following the coup of 1 February 2021.[3]

The subtle play between beliefs and strategy can give us insights into both the successes and failures of rebellion. To understand this, consider a tyrant who has fallen out of favour with a citizenry raring for change.

It is not difficult to see why such widespread opposition to the government at times manifests in open revolt. For that to happen, the need is for a galvanizing incident, such as the self-immolation of Tunisia's Mohamed Bouazizi, a street vendor who set himself on fire on 17 December 2010, thereby creating a focal point for the Jasmine Revolution. Alternatively, there may be a 'focal leader' who can help the citizenry coordinate its actions.

Digital technology has facilitated such coordination. No one wants to go out and protest alone; it is too dangerous because the authoritarian state can have the person arrested, incarcerated or executed. But if thousands of people go out at the same time to protest, each of them is relatively safe, since there are limits to how many people a tyrant can arrest and hurt. The need, therefore, is to coordinate on the time and on the place, which is the classic focal point problem.[4]

We saw the power of new technology in achieving this in Tunisia's Jasmine Revolution. People could exchange messages and make sure that they would not be caught protesting alone or in small numbers. This was made possible by an exile, Amira Yahyaoui. In 2005, Yahyaoui, who was then a young political dissident, had been beaten up by the secret police of the Tunisian leader, Ben Ali, and sent into exile in France. From there she began to help coordinate dissidents using social media. In 2010 she organized an event via the web whereby Tunisian activists took to the streets en masse, simultaneously. The protests gathered momentum and, on 14 January 2011, Ben Ali fled the country, taking refuge in Saudi Arabia.[5]

This was a happy ending, but there are examples of rebellions that were scuppered on the verge of effecting change. Belarus's opposition leader, Sviatlana Tsikhanouskaya, fled to the relative safety of exile, in Lithuania and Poland, and tried to galvanize opposition in the aftermath of the country's 2020 presidential election. This did build up for a while, but the streets of Minsk have since been deserted by protestors.

This gives rise to fascinating questions about why some uprisings succeed and some fail, and which methods tyrants, deliberately or unwittingly, use to foil coordinated rebellions. Perhaps the failure to predict rebellions is because of innately incalculable risks. We may never have an answer to that, but I want to show here how large uncertainties can emerge purely out of the complexity of group dynamics. The hope is that a better understanding of the modus operandi of oppressors will enable us, ordinary citizens, to construct laws and conventions to foil such oppression.

So here is the parable. Consider a leader of a nation, who may have once been popular with the masses but has turned into a tyrant, incarcerating and killing anyone who opposes him. Suppose this nation has a population of 1,000 adult citizens, plus the tyrant and his few henchmen in the police. All citizens want to throw the tyrant out of power. Let's assume that if half the population – in this case, 500 or more people – comes out and joins a protest, the leader will be deposed. Suppose that the citizens have agreed about the date of the protest. The focal point has been created. Each citizen has to choose between joining the protest or being silent. Assume that, unless it is a certainty that one will be jailed for protesting, every citizen prefers to protest rather than be silent.[6]

This is a large country and the tyrant, even if he so wishes, does not have the capacity to arrest and jail so many people. To keep the analysis simple, assume that he has the capacity to incarcerate at most 100 persons. So the situation looks hopeless for the tyrant. If all the citizens come out at the same time to protest, the tyrant's threat to jail people will not work. The probability of a protestor being arrested will be 1/10. That is not enough to deter anybody. So they will protest and the leader will be out of power, as it appeared for a while when the vast majority of Belarus seemed ready to throw out Alexander Lukashenko.

Of course, Lukashenko – I mean, the tyrant – can stop 100 people by naming 100 individuals and announcing that these 100 individuals will be arrested and incarcerated if they are seen protesting. This will make the probability of these people being arrested 1 and they will therefore not go out to protest. However, that will not stop the revolution. In fact, the remaining 900 will feel extra safe (since the jail capacity is already accounted for). The 900 will go out and protest and that will be the end of the tyrant.

However, if the dictator is sufficiently intelligent or lucky, he can stop the revolution totally. This is what he has to do. Break up the citizens into 10 groups of 100 each and give them a label. Group 1 consists of 100 opposition leaders, Group 2 consists of the 100 newspaper editors in the nation, Group 3 consists of the trade union leaders, and so on, all the way to Group 10, which is the 100 teenagers.

Then he has to announce – making sure that the announcement becomes common knowledge in society – his plan of arresting protestors. If there are people protesting against his regime, he will ask his henchmen to arrest 100 opposition leaders. If there are fewer than 100 opposition leaders protesting against his government, they will arrest newspaper editors. If there are fewer than 100 opposition leaders and editors protesting, they will turn to Group 3, and so on. They will stop when they have arrested 100 people and the jails are full. Basically, the idea is first to turn to Group 1, then move down the sequence, to Group 2, 3 . . . all the way to 10. What the tyrant has designed is the Incarceration Game (Basu, 2022b).

It is easy to see that once this arrest plan becomes common knowledge, no one will protest. First, all citizens belonging to Group 1 will quickly be silenced. They know that if they are out there protesting, they will inevitably be thrown in jail, and they will not show up. Since every citizen will be able to deduce this, the members of Group 2 will soon realize that if they are out protesting, they will be arrested for sure, since there will be no one from Group 1 protesting. Group 3, knowing that Group 1 will not join the protest, and also knowing that Group 2 will not join the protest knowing that Group 1 will not join the protest, will not join the protest. This inexorable logic of backward induction will carry all the way down to the 100 teenagers in Group 10, who will soon realize that if they are out there protesting, they will be picked up by the police. The streets of Belarus will be empty, and it will appear that nobody opposes the regime of Lukashenko.

The reasoning that was used to create the Incarceration Game is familiar territory in analytic philosophy because it occurs in the well-known Surprise Test paradox.[7] Here is a short retelling of the paradox. The headmaster of a school walks into a classroom and tells the students that the following week there will be a 'surprise test'. The students go home feeling downcast since a surprise exam is never fun. You have no idea when it will happen. Then one student, the class topper, begins to think which day it *could* happen, and she reaches an interesting conclusion. It cannot happen on Friday. Since that is the last day of the week, if it were scheduled for that day, everybody would know at the end of school on Thursday that the exam will be held on Friday. So a 'surprise test' on Friday is a logical impossibility. But if Friday is ruled out, then Thursday has to be ruled out, because if it were scheduled for Thursday, everybody would know on Wednesday that the exam will be on Thursday. But then it would not be a surprise. By backward induction we can carry this logic all the way to Monday, thereby reaching the paradoxical conclusion that surprise tests are not possible.

It would be folly to treat these kinds of common-knowledge reasoning as purely academic exercises.[8] As the above parable shows, dictators do adopt such tactics, often without full understanding of how they work, and they often succeed. We need to understand the role of this kind of strategic thinking and the power of layered knowledge because authoritarian oppression is common in the world and, from all accounts, on the rise.[9]

In any real nation with a reasonable-sized population, it is difficult to make these kinds of rules of arrest, incarceration or execution – namely, who will be caught first, who second, and so on – common knowledge. It is no surprise that many dictators fail and many revolts succeed. However, there are many examples of tyrannical leaders, from Lukashenko and Daniel Ortega to Vladimir Putin, who have successfully suppressed dissent.

One reason this is not so difficult to do is that, in the problem described above, common knowledge is a sufficient condition but not a necessary one.[10] Rebellions can be foiled with less than that. Much depends in reality on how close we can get to common knowledge. A clever leader can devise ways of spreading the news of her strategy so that the layers of knowledge build up in ways sufficient to foil the protest. Thus, if people would not want to protest if the probability of arrest were high, and not necessarily all the way to 1 – which is what would be the case in reality – then even if you are not certain of being arrested, if the probability is high, you will not protest.

There are, of course, complications in reality. The Incarceration Game involves peering into other people's heads and there is no fool proof way of doing so. Hence, these methods of quelling dissent could fail. Much will depend on the intuition and cunning of the authoritarian leader. Further, there are individuals, such as Mahatma Gandhi, Václav Havel, Nelson Mandela, and Martin Luther King, whose moral commitment is so deep as to make them 'irrational', in the sense in which we use the term in game theory. It may not be possible to get such persons to modify their behaviour through threats.

Often, change comes from the presence of such people. As Bernard Shaw put it more colourfully, in *Man and Superman*, 'The reasonable man adapts himself to the world: the unreasonable one persists in trying to adapt the world to himself. Therefore, all progress depends on the unreasonable man.'

Having a moral compass can be valuable for society. However, as Greta's Dilemma highlighted, it is not a sure-fire solution. Good people in strategic environments can end up doing harm they never intended. My aim here is not to solve the problem (no doubt from the awareness that I am unlikely to succeed) but to show how close to reality we get with moral philosophy and game theory.[11] The task of solving the problem, in reality, is hard.[12] Just as it is not easy for dictators to foil rebellion, because they cannot design the punishment scheme and publicize it enough for it to be common knowledge, or they fail to deal with the few irrationally moral persons in society, we may not succeed in creating prior agreements and constitutions to restrict the power of the leader. Still, the Incarceration Game, drawing on the Surprise Test paradox, lays out the kind of challenge we have to try to meet.

For a better world, beyond the game of life

The moral intention is necessary, but not enough. It is the combination of moral intention and scientific analysis that creates hope.

Tragedy occurred in South Korea on the night of 29 October 2022 in Itaewon, in the heart of Seoul, where more than 100,000 people, mainly teenagers and youngsters, had gathered to celebrate Halloween, to celebrate life. As the crowd built up, it slowly swelled to the point where people were packed in, triggering panic. People tried to rush out of it, making matters worse as they were unable to move out of the narrow lanes that were blocked with people trampling over one another. More than 150 people died from suffocation, a tragedy of staggering proportion.

In the aftermath, reporters struggled to explain what exactly had gone wrong. The tragedy did not have a natural cause, like an earthquake or a storm. Nor did it have a perpetrator, like a dictator giving orders to encircle and shoot. Not being able to place the blame for a human tragedy leaves people feeling uncomfortable, almost complicit. However, we have seen time and again that there are collective outcomes where we cannot point a finger at some individual or even group who bears responsibility.

Does this mean we have to be reconciled to these tragedies as hapless observers? The answer is no – or, more accurately, hopefully not. The clue to a better world lies in looking into the future, anticipating where problems can arise and taking prior steps to prevent them. For natural calamities, like earthquakes and tsunamis, this involves the natural sciences. We need the intention and also physics and chemistry. For human tragedies of collective action, such as war, conflict, oppression, certain kinds of climate calamities, and what happened that fateful evening in Itaewon, we need the social sciences, game theory and mathematics, in addition to the moral intention. Ideally, we want to plan the future from behind what the philosopher John Rawls (1971) called the 'veil of ignorance'. That is, it should be a society I would still want to be part of without knowing what my social position would be. This ensures we are impartial while planning action.

This is the reason for having a constitution or a manifesto. When leaders and lawyers think of a constitution, when ordinary people and activists draft a manifesto, they think of the future, the few essential rules that will apply for the next hundred years or even hundreds of years. Unlike a new law concerning income tax, which is expected to become effective next month or next year and we have a reasonable idea of how it will affect each of us individually, a constitutional provision is typically a broad rule for the distant future. My own identity gets blurry in such a scenario. Hence, when we think about the constitution, there is a *natural* propensity to do so from behind the veil of ignorance. When drafting a manifesto or a constitution, most of the time one knows that it will take a long time for it to become effective and, once that happens, the hope is that its articles and agreements will last a long time. The identity of the self is not zero, as it should be behind the veil of ignorance, but hopefully not salient either, as is the case when we strategize and act in everyday life.

Today, the ground beneath our feet is shifting. Dangerous forms of polarization are giving rise to authoritarianism. It is folly to think of all our existential risks as arising from natural processes and calamities, as happened with the dinosaurs. We can be the cause of our own extinction. The responsibility is a collective one: each person going about their daily chores, working in their own interest, may not result this time in Adam Smith's invisible hand conferring well-being on all. It may be more like the invisible hand of malice, where no *one* has responsibility but we end up creating a sinister society, where minions do their part, with no seeming central authority.[13]

The only way to navigate this turning point for the world is through pre-emptive collective action. This is the time for manifestos and agendas, for agreements and constitutions to try to create a road map for a better world. The closing chapter of this book addresses some of these normative matters explicitly and from a practical, doable point of view.

However, in keeping with the spirit of this book, I want to close the present chapter with two open analytical challenges that straddle game theory and moral philosophy, and urge us to reflect and reason, even if we are not immediately able to solve them.

I have talked in several places in this book about empathy and other-regarding preferences as ingredients for a better world. Standard game theory has a handicap in modelling this. As we saw in Chapter 2, the description of a game includes for each player a 'payoff function', which specifies the payoff a player gets from every possible outcome of the game. This is a given. In game theory, unlike in mainstream neoclassical economics, we have room to allow players to be sensitive to other people's consumption of goods and services. All we need to do is to include other people's consumption of apples, oranges and cars as one of the determinants of my payoff.

However, what traditional game theory does not allow is for my payoff to be dependent on other people's *payoffs*. In reality, a person with another-regarding preference may value an outcome more or less depending on how someone else fares in that outcome in terms of their payoff, compared to how they would have fared in another outcome. I may decide not to choose an action if the other person gets a very low payoff there compared to what she would have got with my choosing a different action. A standard game does not allow for this kind of interdependence.

An exciting research agenda is to create a more sophisticated game theory where each player's payoff function depends on other people's payoff functions. In such a 'sophisticated game', an equilibrium will require first locating a vector of payoff functions, one for each player, such that they are mutually compatible – in the sense that, given the payoff functions of others, each player would have precisely the payoff function that is assigned to her. Then, given this vector of payoff functions, we have to find actions for all players which constitute a

Nash equilibrium, in the usual sense. This is a research problem that can, I believe, be fully solved and, if that is done, it will give us a richer framework for analysing the kinds of outcomes we can expect in societies with empathy and other-regarding preferences. It will also help open up scope for research and analysis for creating agreements and conventions for a better, kinder world.

The second problem I want briefly to comment on is more open-ended. It concerns the idea of the game of life. When we talk of changing the rules of the game, as I have done above, we are clearly stepping outside the game with which we began. After we describe the Stag Hunt game ending up in the bad equilibrium, where individuals fail to coordinate on a specific strategy, if we start to talk about how we can behave differently and get to a better equilibrium, we are stepping beyond the game that was described. In the Stag Hunt, each player could choose between S and H. That they could talk and, by virtue of that, get pre-committed to certain behaviour was not a part of the game. Is it, in that case, at all meaningful to solve the problem of bad outcome by analysing how players should talk before they play? When the game of life is everything there is in the world, what does it mean to step out of the game of life?

To understand this, we need to realize that, in reality, there is no such thing as the game of life. Just as there is no such thing as the 'set of everything', so we cannot describe a game by waving our hands and saying that everything that is possible by the laws of nature is open to a player. This means that there is no natural, well-defined game of life arising from the laws of nature. Instead, the game of life, if we want to use it, has to be a deliberate construct, an agreement among the game theorists. The game of life is for game theorists what a constitution is for citizens. It is an agreement about the boundaries of our analysis, beyond which we agree not to step.

If we take this view, as I believe we *have to*, the problem arises when, seeing a bad collective outcome, we wish to discuss how the outcome can be improved: maybe there should be a conversation among the players so that they get to know each other, we argue; maybe we should encourage the players to have an agreement in advance about how they will play, and so on. The trouble is, this amounts to a flouting of our own convention that there was nothing beyond the original game. How can we have a conversation when our game of life did not have space for a conversation?

It is not clear how we can solve this dilemma. I have to leave this as an open problem. For now, we have to live with reneging on our agreement. That is, we begin by describing the game of life and then, when we do not like the outcome, we violate the terms of the contract among the analysts, and look beyond the game to agreements, conversations and constitutions to help us create a better world.

The important takeaway is that the moral intention in human beings, which I described as so valuable, may not be enough in itself to create a better world. As we saw with games like Greta's Dilemma and the Samaritan's Curse, the moral intention, within the confines of a game, can even make matters worse. The moral intention is useful because it inspires us to step beyond the game and talk and campaign and change the rules and reach long-term agreements. The moral intention is the impetus behind that push for a better world. That is the concern of the closing chapter.

Before crossing over to normative matters, I want to draw attention to a larger issue in positive analysis that opens up once we recognize the essential open-endedness of games. This has implications for everyday life. Game theorists analyse conflict, across groups, nations and family members, in terms of the boundaries of the game. In reality, there are so many more actions and interpretations that occur beyond the visible. Consider messages exchanged between agents, be they heads of states talking about war or householders resolving domestic conflict. So many of our misunderstandings occur – often exacerbating the conflict – not because of what we say or do but because of what is read between the lines or the implicit hints that our actions convey. We convey messages not just by the words we speak but by our silences and what we omit between the spoken words. And what is omitted, be they words or acts, can never be fully defined in real life, since there is no such thing as the game of life and no such thing as the set of everything.

The hidden assumptions that we carry in our heads unwittingly serve us fine till some underlying change compels us to recognize that there were assumptions in our model of which we were not aware. Some of the deepest breakthroughs in research in economics and politics and indeed science occur when we try to recreate and broaden our models, recognizing the presence of these hidden assumptions. It is, for instance, a standard view of the economy from the time of Adam Smith that buyers and sellers, by virtue of prices moving up and down, ensure that demand equals supply for all goods and services. However, this implicit assumption of buyers and sellers facing the same price that is automatically determined is beginning to get frayed. With the arrival of new technology and big data, there is an increasing incidence of bilateral trading, where prices vary from one buyer to another, and also across sellers. Alongside the functioning of markets, we exchange favours that complicate these models. As corporations learn more about your preferences, or about how patient or impatient you are with searching, they can vary the price you pay. Technically, this may still allow markets to be efficient, but at the cost of large segments of the population being exploited totally, with levels of inequality we find hard to fathom. To understand and model this, we have to step beyond normal science to interrogate the fundamental truths in our societies.

7

Manifesto for a Better World

The human predicament

Clouds have gathered over our horizon, clouds that cast ominous shadows of a deteriorating environment, of conflict and chaos, of inequality and injustice. A spectre is indeed haunting the world. It is reminiscent of W. H. Auden's evocative lines, penned sitting in New York's Dizzy's Club, on the eve of the First World War:

> Uncertain and afraid
> As the clever hopes expire
> Of a low dishonest decade[1]

As globalization gains pace, with goods and capital flowing across nations, human prospects have increasingly come under a common umbrella, with shared concerns as well as shared hope. Maybe for that very reason, globalization pits individuals and groups against one another in ways rarely seen in history. Our epoch of changing industrial relations, of diminishing demand for labour, of cabals and collusion, is, for many ordinary citizens, a time of anxiety and despondency, prompting many to pull down the shutters, call it a day and retreat.

For us human beings this may feel like the existential moment the dinosaurs faced sixty-six million years ago. They too may have seen clouds gather on the horizon, but they were mute spectators who had no volition and would soon be extinct, becoming fossils for human beings to dig out, dissect and display in museums millions of years later.

We have one advantage the dinosaurs did not have – the ability to introspect, analyse and change. Therein lies hope, the hope of reason and collective action, not just to weather the storm, but to come out of it better, creating a world that is more equitable, with prosperity shared across regions and across generations. For that reason, we must not give in to the urge to pull down the shutters. We have to remind ourselves that even the darkest of winter nights has a radiance. Given the recent trajectory of history, if we do nothing, there is a lot to lose. On the other hand, if we use our reason, rise and reform, we have a world to win.

We have seen examples before of effective collective action, where people got together to deliberately steer the course of history. The French Revolution of 1789, the American Civil War that began in 1861, the Russian Revolution of 1917, India's non-violent revolution to shake off the yoke of colonialism, which culminated in the nation's independence in 1947, and that fateful day of 11 February 1990, when Nelson Mandela walked out of prison after twenty-seven years of protracted struggle against apartheid and racial oppression, are all examples of this.

What makes these *human* uprisings special is that, in the very end, they are rooted in nothing else but words: debates, speeches and manifestos. The writings of the Enlightenment philosophers of Scotland, England and continental Europe paved the way for the French Revolution. Abraham Lincoln's rousing speeches against the injustices of slavery and apartheid provoked the Civil War, with alt-right groups in the Confederate states worried that Lincoln's victory meant they would lose their 'right' to exploit. The writings of Karl Marx inspired the Russian Revolution, and influenced Nelson Mandela, who was a self-proclaimed socialist and also a member of the South African Communist Party for a while. India's war of independence had roots in associations like the Anushilan Samiti, a revolutionary group in Bengal that made plans for the nation's freedom behind the facade of a body-building club, and later in the prodigious writings of Jawaharlal Nehru, Mahatma Gandhi and Rabindranath Tagore.

No codification of the ideals of revolution was, however, more influential than *The Communist Manifesto* of Karl Marx and Friedrich Engels, published in 1848, in the midst of Europe's republican revolutions. It would later inspire not just the revolution in Russia but the establishment of the People's Republic of China in 1949, the Cuban revolution of Che Guevara and Fidel Castro, and much more, with reverberations around the globe. It is not difficult to see its appeal, rooted as it was in the theoretical writings of Marx, and in its powerful normative yearning, best summed up in the line, 'From each according to his ability, to each according to his needs.'[2]

This line, with its moral resonance and its appeal to human

compassion and kindness, could have been a quote from a holy text. It could be a line taught by preachers and spiritual leaders. It is no surprise that we find similar sentiments in the writings of Thomas Aquinas, the thirteenth-century Dominican friar and priest. In his classic *Summa Theologica*, he wrote, 'In cases of need, all things are common property, so that there would seem to be no sin in taking another's property . . . [I]f the need be so manifest and urgent that it is evident that the present need must be remedied by whatever means be at hand, then it is lawful for a man to succour his own need by means of another's property, by taking it either openly or secretly . . .'[3] No wonder some of our most radical ideas have their origins in theological writings.

However, history did not go by the scripted path. Somewhere along the way, Russia's Communist revolution morphed into one of the worst forms of crony capitalism and oppressive oligarchy. The early idealism of China's revolution gave way to brutal authoritarianism and dictatorship. Even in smaller countries we have seen a similar trajectory. In Nicaragua, for instance, Daniel Ortega, imbued with Marxist ideology, overthrew the evil and corrupt regime of Anastasio Somoza Debayle and came to power, only to turn into a dictator himself, surrounded by sycophants and flatterers.[4]

The divergence between what many progressive movements aspired to and where they ended up has had an unfortunate consequence. Each failure is used by ultra-conservative groups and those with vested interest in the current iniquities to stall all progressive initiatives to redistribute income and wealth. By equating the aim to achieve greater equality and banish poverty with the cronyism and tyranny that evolved in

Russia or Nicaragua, they have tried to thwart all progressive movements.

We need to recognize that the aim of a fair and compassionate world is admirable, even though Marx's blueprint of how to get to that state was deeply flawed. In other words, Marx got his normative ambition right, but positive economics wrong. The person who sensed this mistake, though he wrote in allegories rather than as a social scientist, was Franz Kafka. As he lyrically described in his novel *The Castle*, one may be right to yearn to get away from this humdrum life to a shimmering castle, but one must, at the same time, be sceptical about whether there is a viable path to it.

Peering at the world today, it seems likely that there is turbulence ahead, because of tectonic shifts beneath our economy and society, caused in part by our own folly and excesses of exploitation and indulgence, but also for natural and almost inevitable reasons.[5] If we treat the turbulence not as a time for resignation but as an impetus to action, we may actually succeed in parting the clouds, letting in the sun and settling on a better trajectory.

The mistake that idealists and radicals made in the past was to ignore the incentives of individuals in designing a system. We have to recognize that the drive to do well for oneself and the profit motive are part and parcel of *Homo economicus*. This is not to deny that this may change in the future. Norms change and we make progress. In some societies, you can walk down a lonely street at night, wallet bulging, and feel no anxiety that it will be taken away from you. In other societies, there is a reasonable chance that it will be gone, even without too much of a bulge. In some societies, you can stand in a queue leaving space between yourself and the person in front of you. In other societies, there will be no reason for the neoclassical economist to be disappointed, because it is likely that someone will maximize utility and slide into the gap. We worry about the village pond being overfished because of human selfishness and the commons problem. On the other hand, the household refrigerator is typically not kept locked, but it does not get drained of all food supplies. We do not eat all our favourite food because of the awareness that an old aunt needs it more. Our norms and moral wiring help to solve the commons problem of the household.[6]

It is conceivable that a time will come when we can move beyond small groups, like family or household, to the level of the nation or society at large, that we can create a world where we contribute according to our ability and consume according to our need, as we so often do in households. However, the world is nowhere near that state. To believe otherwise and try to create a utopian state overnight, where individuals do not have rights to property, there are no private firms and all wealth is centralized and placed in the hand of the state, would be folly. With all wealth in one place, the state becomes a perfect target for cronies. Once they capture the state, we have the worst form of crony capitalism. It is no surprise that history has shown us that crony capitalism is the last stage of Communism. Russia stands testimony to this.

This realism must not be treated as reason for accepting the status quo, nor for the even greater folly, seen in many conservative circles, of treating the status quo as ideal. The balancing act between an all-embracing state and laissez-faire is analysed well by Acemoglu and Robinson (2019, pp. 464–7) in a section entitled 'Hayek's Mistake'. Friedrich Hayek (1944), the Austrian intellectual and economist, took his fear of an all-encompassing state and life under an over-arching leviathan to the other extreme of trying 'to rein it back completely'. What Hayek did not foresee is that an all-encompassing state and extreme laissez-faire, while apparently opposite approaches to government, have a similar end result: both converge on crony capitalism. The key to striking a balance is to restrict the size of the state, while having the state play the role of *transferring* wealth from the rich to the poor.[7] As the next section tries to argue, it is possible to do so without shrinking the size of the overall economy and even while nurturing its growth.

We have just to open our eyes to see the intolerable levels of inequality we have in the world today.[8] One does not need reams of data to realize this but just the ability to observe. It is debatable whether the inequality today is greater than before, but what should be beyond dispute is that it is at a level unacceptable by any reasonable normative yardstick. We should be ready to admit this whether we are one of society's winners or losers. The occasional disquiet about the status quo recorded even among the affluent is reason for hope.[9]

Around the world, the effects of alarmingly high economic inequality are spilling over into politics and society. Economic insecurity is a driving force behind violent conflicts in the Middle East and the rise of fascist sentiments in several European countries. Even in old democracies such as the United States, economic marginalization has led to a strengthening of chauvinist and supremacist identities and other social problems.

The World Bank's estimates show that in 2018 (I choose a time before the pandemic), there were 659 million people living on an income of less than $1.90 a day. This is a shocking figure for a world where there are individuals who have wealth greater than $100 billion, implying an income of approximately $14 million a day.[10]

Powerful voices in both rich and developing countries – and, tragically, even some of those who stand to lose most from the current system – claim that our income disparities are fair because they are the result of free markets, which are needed to create incentives. The right incentives create growth, it is argued, and that growth in overall wealth will eventually 'trickle down' to those who need it most. It is true that *some* inequality is needed to create incentives, but not the inequality of a world where some people have an income of $13,690,000 a day and over 600 million people live on $1.90 a day.[11]

I am convinced that there will come a day when human beings, looking back at our times, will wonder how we tolerated this kind of inequality. They will be shocked by our complacency, the same way that we are shocked at the way our ancestors tolerated slavery and the selling and auctioning of human beings, or the caste system and the practice of untouchability. The argument that today's inequality is fair because it reflects choice – the choice by some to work hard and take risks, and the choice by others to have less money but more leisure – is shattered as soon as you recognize the simple fact that the bulk of the poor are born poor. Since babies do not work, wealth at infancy cannot be a reflection of one's preference for hard work. From theoretical research to empirical studies, we now have ample evidence that poverty acts as an intergenerational trap.[12] It follows that much of today's inequality is arbitrary. Even if we can justify other forms of inequality, we cannot justify some being born destitute, and some rich beyond imagination, on the grounds that this is fair.

To rectify this and move towards a fairer and more equitable world, without damaging incentives and ending up creating an even worse scenario, we first have to recognize that human beings will have a certain amount of their own interest in mind. We also have to acquire an understanding of how markets function. The lessons of economic theory and game theory, such as the themes discussed in this book, are germane to our task. We have to let private enterprise flourish and markets function without excessive bureaucratic interference. We do not want government to amass great wealth, which invariably leads to capture and cronyism. This means that the state must not hold too much wealth of its own but instead play the role of transferring wealth from rich to poor.

Can all this be done without damaging people's incentive to work hard and create wealth for themselves? I believe the answer is yes. Yes, if we are willing to begin with facts and use our reason. I began this section with words of despair from Auden. In closing the section, and before going on to outline the need for activism, let us remember the defiance of the same poem, which reminds us that we have

> [. . .] a voice
> To undo the folded lie,
> [. . .] And the lie of Authority
> Whose buildings grope the sky

The accordion tax

The richest person is richer than the poorest not ten times, a hundred times or a thousand times, but somewhere in the vicinity of ten million times. Surely this is unacceptable. Apart from anything else, it damages democracy because, with this kind of staggering income gap, the rich have a thousand ways to silence the poor. The poor are left with their franchise but with little else, and certainly with no effective voice. With so much money concentrated in the hands of the few, the richest people will influence public opinion, buy up politicians and silence dissent. Once that happens, it will be impossible to turn things around for the circular reason that the majority has lost its voice.

A related phenomenon, which is associated with poverty and extreme inequality but received less attention because we did not have the vocabulary for it, is the disparagement of the poor per se. Adela Cortina (2022) coined the term 'aporophobia' to capture this phenomenon, and discusses how painful aporophobia can be. There is nowadays talk on discrimination based on race and gender, for instance, but little on discrimination against the poor, when there is no other associative identity. We need to address extreme inequality even when it is not associated with other markers.

Some of these problems are getting more acute because of advances in digital technology. An example is the digital platform for buying and selling products and services, such as Amazon, Uber and Airbnb. We tend to analyse this new economy using traditional models of economics, where individuals are utility-maximizing cogs in a giant machine, but that model no longer serves us well.[13] In traditional economics, buyers and sellers were brought together by the invisible hand of the market or the imaginary auctioneer. With the arrival of the digital platform, the auctioneer is no longer imaginary but an actual corporation, with an eye on little other than its own profit. This is causing large concentrations of wealth at the top and nudging us to think of markets in novel ways. Further, with the advance of digital technology, we are also forced to think of data and information as property, thereby making us think of laws and policy in novel ways.[14]

Traditional laws, such as those pertaining to antitrust, are proving to be inadequate. It has rightly been criticized that these laws are too focused on the consumer, and we need to turn attention to small retail traders and also labourers.[15] However, even beyond this, a problem arises from the fact that the advantage of these platforms is their size. We cannot rectify the problem by using antitrust law and breaking up the platforms into many firms. Some of them cannot simply be broken into parts and, if they were, much of their efficiency would be lost. What, then, can we do? One option is to let these firms earn large profits but make sure that the shareholding of these firms is widely dispersed, thereby ensuring that the large profits go into a large number of pockets.

In the case of some of the *largest* and *most critical* platforms, we may need to think of more radical policies, such as declaring them to be non-profit. When a firm becomes the gateway to the whole economy, so that everybody has to enter this gateway to survive, we cannot leave that firm in the hands of the private sector. It is often forgotten that the Bank of England began in 1694 as a profit-making entity, with 1,200 individuals holding shares and earning dividends. However, gradually, it became evident that money was the gateway to the modern economy. No one can live by barter any more. This led to the realization that money creation cannot be left to a profit-making corporation. Eventually, central banks would become non-profit entities. The time has come when we may have to think along similar lines for some critical digital platforms.[16]

There are other interesting ideas that have been floated to rectify the worst inequities. A particularly popular one is that of a universal basic income, which is a state guarantee that everybody will get a basic minimum income. In the absence of other corrective measures, a universal basic income is attractive. However, it is less than ideal. It is focused entirely on extreme poverty alleviation and pays too little attention to inequality, which creates its own problems.

We must also keep in mind the pragmatic consideration that a flat guaranteed income can damage the incentive to work and be creative. For the same reason, we have to be cautious with the idea of an income cap. Inequality can be controlled if we place a cap on the highest income people can have. But as more people hit this ceiling, this will have a severe dampening effect on the incentive to work hard and innovate. Not only can you not earn more by working hard, you cannot even improve your *relative standing* in society. Fortunately, there are ways to tax and redistribute income substantially while doing minimal damage to individuals' incentive to work.[17]

It is reasonable to assume that, *once basic needs are met*, the major reason for people's hard work and enterprise is to promote their *relative* well-being, that is, where they stand vis-à-vis others. Mainstream economics assumes human beings work to maximize their absolute well-being, typically measured by their income or wealth. However, this needs a lot of qualification.

First of all, there are professions in which people are willing to work no matter what the financial return. This is mostly true of the arts and the pursuit of knowledge. Picasso's passion was painting and he would, in all likelihood, have painted even if he had earned nothing by that. It so happens he earned a lot. Van Gogh's passion was painting and he painted, even though he earned nothing from that. Then there was Socrates. In the words of Bettany Hughes, 'Socrates spun through Athena's city like a tornado . . . Women, slaves, generals, purveyors of sweet and bitter perfumes – he involved all in his dialogues.'[18] Xenophon, a student of Socrates, remembered him spending the entire day in the marketplace talking to anyone who would care to listen. In *Euthyphro*, Plato quotes Socrates as saying: 'Because of my love of people they think that I not only pour myself out copiously to anyone and everyone without payment, but that I would even pay something myself if anyone would listen to me.'[19]

Moving away from van Gogh, Picasso, Socrates and a host of other creative people, we find that even people who work to earn money, once they have crossed a certain basic income and their essential needs are met, work to earn more primarily to enhance their *relative* income. The enormous effort that well-off people put into being rich is not in order to have access to the third yacht but to *show* people their third yacht. There is a lot of literature on the importance of *relative* payoff goals from at least the time of Thorstein Veblen (1899). When people are poor, the primal urge is to survive and thereafter to reach a certain basic standard of comfort, but once these basic needs are met, the primal need shifts to relative performance – doing better than the neighbour next door or, in today's digital age, our Facebook friends.

This gives rise to opportunities for policy interventions that can create a vastly more equitable society without damaging the urge to work. The instrument I would like to propose for achieving this is the 'accordion tax': a system of taxation which flattens out the income distribution but does not change the rankings of incomes. In other words, we can leave the drivers of national income growth largely intact, while flattening out the great inequalities. Further, the proposal I am about to put forward does not require a large government, with centralization of wealth.

To describe an accordion tax, we have to select a cut-off income level, which will be called the 'hinge' and has to be greater than or equal to the average income of society. Then we have to choose a percentage tax rate, *t*, which is less than 100 per cent. The hinge and the tax rate fully describe an accordion tax system. All individuals earning higher than the hinge are taxed *t* per cent of what they earn above the hinge. Thus, if the hinge is at $50,000, Amanda has an income of $62,000 and the tax rate is 50 per cent, then Amanda has to pay a tax of $6,000. The money collected under this rule is transferred to people with income less than the mean income. The poorer the person, the higher the transfer they receive. This is the broad idea of an accordion tax. The exact rule for the transfer of money to those below the mean income is explained below.

This is not a part of the formal definition of accordion tax, but the idea is that if the mean income is above the level where basic needs are met, the hinge should be set at the mean. If the mean income is below the level of basic needs – that is, we are dealing with a very poor country – then the hinge should be set at a higher level, in particular, some place where the basic needs are met. Since above this level the hard work is for relative higher income, it will be possible to raise the tax rate without damaging the incentive to work.

It is useful to describe formally a special case of accordion tax, which I refer to as the 'strict accordion tax'. Assume that society's mean income is higher than the basic needs and the hinge is set at the mean income. People with income higher than the hinge are taxed as described above. And each person below the hinge is given an income subsidy equal to *t* per cent of the gap between the mean income of society and the person's income. It is obvious that the entire money collected from those who earn above the mean will get handed over to those who earn less than the mean. In fact, the easy way to describe the strict accordion tax is that everybody has to pay a tax which is *t* per cent of the gap between the person's income and society's mean income. Since this gap is negative for those below the mean, people who earn less than the mean pay a negative tax, that is, they *receive* money.

Unlike a universal basic income, the accordion tax leaves income rankings unchanged for everyone – the top earner is still the top earner. A tax rate even as high as 99 per cent will not change any rankings but flatten out the income distribution greatly.[20] Since it is relative income that largely drives people (once above a basic income level), the work incentive will be affected minimally by this tax. Elon Musk will not stop working, nor will Mrs Jones next door. Musk will want to keep the lead over Jeff Bezos, and Mrs Jones over Mr Smith.

Suppose we have a society where the per capita income is $50,000, the poorest person earns $500 and the richest person earns $5,000 million. So the richest person is 10 million times richer than the poorest person. If we treat $50,000 as the hinge and impose a tax rate which will make the richest person 10 times richer than the poorest, we have an accordion tax system where a vast amount of the billionaires' income will be taxed away, the poor will be much better off and all rankings will remain unchanged (and the richest will still have the opportunity to brag about being the richest).

Let me now describe the accordion tax formally, without the qualifier 'strict'. An accordion tax is defined by a hinge income, which is greater than or equal to the per capita income, and percentage tax rate, *t*, greater than 0 and less than 100. As seen before, those earning above the hinge pay a tax equal to *t* per cent of the income earned by the person above the hinge. And those with income below the per capita income are paid an income subsidy, which is *s* per cent of the gap between society's mean income and the person's own income. If the mean income is $10,000, the person's income is $6,000 and *s* is 40 per cent, this person gets an income subsidy of $1,600. The *s*, however, is not arbitrarily chosen. It is picked so that the total tax collected from all who earn above the hinge and pay taxes adds up to the total subsidy received by those earning less than the mean. It is easy to see that if the hinge is set at the mean, *s* becomes the same as *t*, and what we have is the strict accordion tax.

The intuition behind the accordion tax is simple. We set a hinge above society's per capita income. Those who earn an income above the hinge pay a tax, and those who earn below the per capita income get a subsidy. Those whose income lies above the mean and below the hinge neither pay a tax nor get a subsidy. The total tax collected by the government is equal to the total subsidy handed out. The accordion tax is fiscally neutral.

In reality, we may want to modify features of this idealized system to suit specific contexts. We may, for instance, want to raise more taxes than we pay out as income subsidies, in order to finance other fiscal needs of the nation.

There is also a practical problem with raising the accordion tax too high in a specific nation. In today's globalized world, if the tax is unilaterally raised very high in any nation or a small cluster of nations, this will likely cause capital flight out of those nations. To prevent this will require global coordination, whereby all major economies agree to use these taxes with some coordination. As we move from considering individual actions to group behaviour, such as prescribed by a global constitution, it is inevitable that we will need to consider collective agreements, an important topic in recent writings on philosophy, politics and economics.[21]

The idea of a global constitution is a many-layered problem; we need underlying institutions to make the commitments credible.[22] The core idea behind a constitution is a set-valued equivalent of the idea of a Nash equilibrium that we encountered in Chapter 2. It is basically an agreement among all agents in which each person agrees to give up certain actions and behaviours, and, given that others are willing to give up what they have agreed to give up, it is not in anyone's interest to violate their own commitment.[23]

It is possible to take this idea further by bringing in the role of public reasoning. In pluralistic societies, public reasoning has long been recognized as an essential tool for agreements.[24] Based on this, it can be argued that constitutions should be formed following public reasoning and deliberation, since such deliberations, even though at one level they are just 'cheap talk', can lead to legitimacy of certain kinds of behaviour, and result in self-enforcement, which would not have been possible without the initial discourse. There are studies showing how players in a game get disutility from violating commitments they have agreed to abide by.[25] There is scope for taking the methodology advocated in this book, namely the formalization of some of the philosophical writings and ideas with game-theoretic instruments, to bolster public policy. The adoption of a constitution for today's increasingly globalized world is one such pressing problem.

To think of making the idea of a constitution work at a global level may seem like a pipe dream. However, we must not give up. Having a constitution for a large and diverse nation like the United States must have seemed impossible, until it happened in 1789. Fortunately, there were far-sighted leaders with intellectual commitment and the aspiration to create – to use Abraham Lincoln's words from the memorable speech in Gettysburg, Pennsylvania, on the afternoon of 19 November 1863 – 'a new nation, conceived in Liberty, and dedicated to the proposition that all men are created equal'.

This dream in turn would have amounted to little if not for ordinary people who were fired with ambition to go beyond their immediate self-interest in order to create a better collective future. To carry such a task to a global level, creating an agreement, however minimal, for all of humanity, is difficult, but not impossible. Our capacity for analysis and the ability to formalize powerful ideas from philosophy with game-theoretic reasoning have grown. It is not surprising that in today's world we have minimal amounts of trade policy coordination via the World Trade Organization and some labour regulation at a global level via the International Labour Organization; we have the International Criminal Court, where we can try leaders of nations for violation of basic human rights; and we have a European Union and a common currency, which would have seemed a pipe dream a few decades ago.

Going further back into history, we have examples of movements like the Chartists, who tried to unite workers and give them a voice by directly appealing to the coordinating effects of a charter.[26] Feargus O'Connor, the Irish maverick leader of the movement, went from door to door collecting signatures of support in the belief that the very act of signing would allow the workers 'to see themselves as a distinct class'. He pointed out that not only would this enable radicals to think alike but to *know* that they thought alike.

It is time to think more aggressively about a global constitution, with minimal global agreements, such as for upholding basic human rights – beyond what is currently being done – and holding leaders responsible for violations, placing term limits on leaders in every country, and having provisions for coordinating fiscal policy and systems of taxation.

This will not happen easily because it will inevitably hurt some vested interests. At the same time, as globalization continues, if we do not succeed in coordinating more policies across nations, we risk exposing ourselves to the threat of extinction. It may not be in the interest of all current inhabitants of the Earth to create a more equitable world, since some are doing well *by virtue of the inequities.* But in this game there are many bystanders who lose out from the current system, and future generations who will inherit the world from us but have no influence on what we do. Our hope must lie in the fact that, throughout history, there have been individuals who, even when they themselves were beneficiaries of the existing system, were willing to speak out against the system. Even if they did not give up their advantage unilaterally, they had the moral courage to strive for systemic change.

Further, those who face the brunt of these systemic injustices must organize and act. If current vested interests try to block change, we will need to create global grassroots movements whose voices will be heard. We go about life's daily chores trying to choose our actions and strategies from the familiar feasible set to maximize our payoffs. It is time now to use our imagination to step beyond the game of life to explore new actions that have been lying dormant thus far. We live in a small world. We need much more aggressive action for justice, equity and fairness, cutting across nations and spanning generations. It is time for placing scientific ideas on the table, and time for action.

Beyond nation

It is time to nurture our *human* identity and strive for equity and justice cutting across nations and reaching people wherever they are. One of the unfortunate trends of our times is hyper-nationalism, the promotion of the narrow view that one's country is the greatest in the world. Apart from the incoherence of each nation being the greatest – possible only if all nations in the world score exactly the same on the greatness index – this is a supremacist view, on a par with racism, which should have no place in a world of equals. There will come a time when chest-thumping about one's nation will appear as morally degenerate as the tales, luckily receding into history, of pride about being white or male or upper caste.

Nationalism did play an important role in history. For people to make sacrifices in order to promote the larger good, we often do need to feel a sense of identity with a group. As India's first prime minister, Jawaharlal Nehru, soon after India's independence, pointed out in one of his periodic letters to the chief ministers of the nation, 'The feeling of nationalism is an enlarging and widening experience for the individual or the nation. More especially, when a country is under foreign domination, nationalism is a strengthening and unifying force.'[27]

There is a lot of work we do as collectives, from family, community and clubs, to firms and corporations. These identities can confer benefits. As the economists Samuel Bowles and Herb Gintis point out, they can help 'overcome free-rider problems and punish "anti-social" actions by supporting behaviours consistent with such pro-social norms as truth-telling, reciprocity and a pre-disposition to cooperate towards common ends'.[28] This leads to efficiency and can be good for productivity.

There are, however, two caveats to keep in mind. First, with globalization, we have reached a stage where we also have to nurture our largest group identity – that of being cohabitants of the Earth – and nurture pro-social behaviour across humanity. Secondly, the sense of identity, including the national one, is something we should feel good about, but there is a distinction between feeling good and feeling proud. To feel good about one's identity is to bolster confidence; to feel proud is to diminish others.[29]

Immediately after the lines quoted above, Nehru, ever mindful of the degradation that comes with hyper-nationalism, went on to remind the chief ministers, 'But a stage comes when [nationalism] might well have a narrowing influence. Sometimes, as in Europe, it becomes aggressive and chauvinistic and wants to impose itself on other countries and other people. Every people suffer from the great delusion that they are the elect and better than all other.' These are remarkable lines coming from the prime minister of a nation, especially keeping in mind that India had just gained independence. It is a rare leader of a nation who reminds his people that nationalism can become 'a narrowing influence'.[30]

There is now research by psychologists that distinguishes between different kinds of moral values, such as those related to caring for others and fairness, and those that involve in-group loyalty and obedience to authority.[31] Studies show that in-group loyalty often conflicts with genuine pro-social behaviour and caring, and can lead to condoning violence towards and exploitation of other groups. Some of these authority-based values can indeed become the basis of what to any objective outsider will appear immoral behaviour. Nationalism runs this risk.

Some of our identities are natural or even biological, such as those pertaining to family or culturally similar groups. There are people with a shared history, who grew up listening to similar music, reading the same books, looking at similar artworks. There is some overlap between these identities and the national identity. However, national identity, which once helped us to strive and do well together, can become an instrument of oppression, used to justify fighting and oppressing other peoples and other nations. This has a long history but reached new heights in the twentieth century with the two World Wars, and is on display in many places today.

Nationalism has functional value and, for that reason, it has survived through long stretches of history. It was not there among the ancient peoples, gradually emerged and strengthened, and helped large groups of people to cooperate and prosper. As, thanks to the steady march of technology, the world gets more globalized, with goods, services, money and people criss-crossing the planet, nationalism runs the risk of becoming more damaging than valuable. There is reason to believe that hyper-nationalism can become a major ingredient of our dissolution. Like racial supremacy, hyper-nationalism was always morally questionable, but now there is the additional problem of its being destructive, not least as an instrument to exploit the poor for the narrow gain of the ruling elite.

We have seen in earlier chapters of this book that what we aspire to – or, in the language of game theory, our payoff functions – are not always hard-wired in us. Instead, new targets and ambitions can be created and nurtured so that people are willing to work hard and even make sacrifices to achieve those targets. This can be a source of hope but, at the same time, combined with hyper-nationalism, it can create a deadly brew. Nowadays, we see nations advertising when their richest person climbs up the chart of the world's rich.

We could envision a scenario in which a rich person in some poor nation starts a crowdfund for people to help him climb further up the global chart, and gullible people contribute to this 'national cause'.

As money and markets criss-cross national boundaries, we have to expand our terrain of concern. It is not enough to go beyond the individual to the group or even the nation. We have to think of the world, and future generations to whom we bequeath the world, as the domain of policy and law. This is the challenge we face today. A large part of globalization, which is the outcome of technological advance, is a part of life. To rail against this is like blaming gravity for the collapse of a building. Gravity is beyond doubt a key factor behind every building that falls, but it is not a cause we can do much about. For that reason, we do not dwell much on it, focusing instead on other causes, like engineering flaws that may have caused the collapse and against which we can take corrective measures. Likewise, for the strains on the economy and polity caused by globalization and the march of technology, we have no choice but to take these as given, and think of measures that are within the domain of our control and can curb the negative fallout of this change. This will require radical and out-of-the-box thinking.

Towards global justice

This book began with the individual, the economist's *Homo economicus*, and with an introduction to game-theoretic reasoning. Game theory has been applied extensively to matters of corporate competition and regulation, and war and diplomacy. It can also be of great value to the individual in the conduct of everyday life. Early philosophers like Socrates and Epicurus insisted and demonstrated by their own lifestyles that philosophy is not just an intellectual pursuit but a way of life. Socrates is, of course, a household name, but Epicurus deserves as much appreciation. Not only was he one of the deepest thinkers, his lifestyle was radical in a quiet way, somewhat akin to David Hume's. He lived simply, had the courage – for right or for wrong – to reject Platonism, the dominant philosophy of his time, and founded his own school, The Garden, where he admitted slaves and women, which was anathema at that stage of history.

Game theory, which is barely one hundred years old, was not available to the Greek thinkers, nor to the European philosophers of the Enlightenment. The opening chapters of the book tried to give shape to some of these ancient intellectual yearnings by bringing game theory to bear on them.

From the pursuit of self-interest, the book proceeded to discuss the moral yearning that is also innate in all human beings, even though many traditional neoclassical economists deny this. This compelled us to move from the individual to the group, such as class, community and even the nation, and investigate a vexing question: can a group of moral human beings exhibit morally reprehensible group behaviour? The answer – this should not have come as a surprise to the reader after having read the opening chapters – turned out to be yes. The agency for collective evil is not always easy to attribute to the individuals comprising the collective. This is an important moral dilemma and raises a host of issues regarding collective morality.

The previous two chapters sounded a warning: that whatever

our morals, we have to be mindful of the unexpected outcomes of group behaviour. This took us into practical matters like regulation and collective agreements, and raised philosophical questions concerning the meaning and existence of the game of life.

Moving from the individual to groups – small ones, like family, and large ones, like nation – led us inevitably to what no analyst today can look away from: the world. With our increasing connectivity, not just through travel but via digital links, globalization has brought people to the same shared space sooner than our intellectual capacity and moral antenna were prepared for. Hence, the unavoidable closing chapter of this book. This chapter gives the reader a glimpse of the problems of rationality, morality and strategy, faced by the largest collectivity, as we know it, at least at this stage of science: the world.

In closing, I want to remind the reader that human civilization is at a juncture where we, the current inhabitants of the Earth, need to think of our *common* human interest – in fairness, in justice, in survival, in sustainability. We have to think not just of our own payoffs in the 'game of life', but of the well-being of bystanders who live on the fringes of society. We also have to think of those who cannot reach out to us, but who will inherit the Earth after we are gone. We live most of our lives without a thought about who these bystanders are, their joys and miseries, and about our collective responsibility for their predicament.

As technology advances in leaps and bounds in a globalized world, there is not much we can do to stall some of the most fundamental processes that move virtually by the laws of nature. There is, however, a lot that we can do if we are willing to draw on our resources of kindness, love and empathy, and marshal our intellectual ability to design new interventions. As it happens, some of this is in our own enlightened interest, which reminds us that we can no longer live insulated lives, taking pride in narrow group identities, sheltering behind the walls of a nation, unmindful of the dangers and the injustices beyond our national boundaries. Suffering elsewhere will wash up on our shores and, even if it did not, we have a moral responsibility to fight against human suffering wherever it occurs. And, for this, it is not enough to have the right moral intention; we also need science and sagacity. As we saw with the case of Karl Marx and Friedrich Engels, some of the best intentions can come to naught, and even come to haunt us if we fail to combine them with reason.

This book is primarily about reasoning, which is the key ingredient of all the sciences. However, we cannot rest with that. We are at a turning point of history. There is a general disquiet in the air. The clouds of climate change are darkening the skies. Authoritarianism is once again on the ascendancy across the globe. With the rise in inequality, which silences the voices of people and damages democracy by trampling over basic human dignity, there is urgent need for action. We have to strive for greater equity and justice urgently, before the rich become so powerful as to persuade the poor that transferring wealth from the rich to the poor is bad for the poor.

According to latest estimates, there are over 650 million people who live in extreme poverty, that is, on less than $1.90 a day. If all the people living in extreme poverty had a common identity, such as that of race or gender, we would be shocked at the prejudice being practised on Earth and demand immediate corrective measures. The fact that the poor and marginalized have little else in common other than being poor and marginalized does not make the situation any better. Perhaps it makes it worse, since the oppressed do not have the comfort of a common identity. Many of the fractures of conflict, authoritarianism and political polarization that we are witnessing today have origins in these deep-rooted injustices and inchoate frustrations.

It is time for activism to bring these injustices to an end and pave the way for better decisions, better policy and, ultimately, a better world. It cannot be left solely to those who live in poverty and bear the worst consequences of globalization and the march of technology to campaign for equality and justice. We, including those who stand to lose if there is a better distribution of wealth and income, must lend our voices to fight injustice and create a better world. It is time to break away from our narrow identities, to think beyond race, class and even nation, and consider the well-being of all on our small planet. If our nation wages an unfair war, I hope we have the courage to stand up for the so-called enemy. If we ever have to choose between nation and justice, I hope we have the courage to stand for justice.

Notes

CHAPTER 1

1 Fortunately, these White House discussions were secretly recorded (only President Kennedy and, it is speculated, his brother, Robert Kennedy, knew they were being recorded). There is a thirty-five-year embargo on such recordings but they are available now (see May and Zelikow, 1997).

2 *The National Cyclopaedia of Useful Knowledge*, Vol. I, London, Charles Knight, 1847, p. 417.

3 One joy you can still indulge is proving the theorem in a *new* way. It has been reported (Ratner, 2009) that there are 371 ways to prove the theorem. I myself indulged in this joy by trying to notch the count up to 372 (see Basu, 2017).

4 *National Geographic*, 193, 6 June 1998, p. 92.

5 I am drawing this from my paper, Basu (2014).

6 This example is also a reminder that the method of randomized control trials (RCTs), which has become popular in economics, must not be treated as a technique for establishing causality. RCTs are indeed the gold standard for describing the population from which the sample is drawn, but they do not establish causality (see Cartwright, 2010; Basu, 2014).

7 There is some interesting research in economics on the *instrumental value* of anger. One person's anger can play an important role in deterring another person from violating

norms or not doing what a 'reasonable person' would do (see Miller and Perry, 2012; R. Akerlof, 2016).

8 Tom Stoppard, *Jumpers* (London: Faber & Faber, 1972), p. 13.

9 The informal test I conducted with students at the Delhi School of Economics in 1995 confirmed this propensity to reason poorly when the words had emotional content (Basu, 2000).

10 See Hume (1740).

11 For a fascinating paper on this subject, see Garry Runciman and Amartya Sen in the journal *Mind* (1965). Let me take this opportunity to give the reader a tale from academe. Publishing in top journals is hard game for researchers, who have to be prepared for many rejections. This paper is one of the earliest attempts to use game theory to understand some conundrums of moral philosophy, in this case the ideas of Jean-Jacques Rousseau. The paper, written in 1959, emerged from Sen's collaboration with Runciman. They submitted it to the editor of *Mind*, the celebrated philosopher Gilbert Ryle, and were delighted when the paper was accepted. Then nothing happened for three years; they wrote to Ryle reminding him, and to make sure he knew what they were talking about, they included a copy of the paper. Clearly, Ryle had forgotten about the paper. Taking it to be a new submission, he had it reviewed again and it was accepted. Hence, this famous paper was not just accepted by *Mind*, but accepted twice.

12 See Tait (1977, p. 202).

13 There is one complication here. It is possible to argue that when, after a dinner, you say, 'It is delicious,' you do not mean what we normally mean by 'It is delicious.' Both parties – the cook and the diner – know that this is just an act of courtesy, like shaking hands. Therefore, statements like this need not be true or false.

14 What is not often recognized is that there is a great similarity between Adam Smith's theory of why markets work and David

Hume's conjecture of what confers authority on leaders. Both explanations lie in the choices made by ordinary individuals (Basu, 2018). It is often assumed that Hume rejected the idea of the legitimacy of government depending on a social contract. However, as Sayre-McCord (2017) points out, this may not be the right interpretation of Hume. Later in the book, I take a similar line concerning the invisible hand of the market and the critical role of morals.

15 This argument can be developed in many ingenious directions, and it is possible to close the argument without bringing in morals (see Myerson, 2004; Samuelson and Stacchetti, 2017). If, for instance, this interaction with the taxi driver is followed by a war of attrition, we may not need norms beyond that of helping select focal points. Though this is logically possible, I believe we do have norms that are virtually hard-wired in us, which makes us, in some situations, choose certain actions without thinking. These actions could have roots in what may be described as an innate 'cooperative motive' (Brennan and Sayre-McCord, 2018). This could mean that it is not quite the norms that are hard-wired but the penchant for following norms. People observe others' behaviour and norms emerge (Hoeft, 2019) but, once in place, they are followed instinctively.

16 It is important to recognize that social norms, based on culture, and moral norms, such as fairness, are often intertwined (Elster, 1989; Platteau, 2000). As Myerson (1991) observed, cultural norms help find a focal point but certain cultures may look to equity and fairness to select the focal point.

17 In his recent book, the economist Robert Frank (2020) derives some of these codes of individual behaviour from collective concerns. Because our own actions have such a minuscule effect on collective outcomes, it would be in our own interest to ignore them. Hence, society's success depends on the ability of the collective to develop the codes of individual behaviour, enforced by taxes, regulations and punishment or social norms.

CHAPTER 2

1 As was recognized by the philosopher David Lewis (1969) and economist Robert Aumann (1976).

2 I used this fictitious tale in an article in the *Scientific American* in 2007 to illustrate some of the paradoxes that underlie game theory.

3 My book (Basu, 2018) is a response to this critique and an exploration of how to rebuild law and economics without this fault line (see also Mailath, Morris and Postlewaite, 2017). There is an earlier literature on the scope for governance without government (see Dixit, 2004).

4 Much of early game theory was good at analysing games as one-time interactions, without enough richness for analysing what is called extensive-form interaction between players, which refers to sequences of moves – I do this and then you do this and then again I decide what to do. A series of papers written through the 1960s, 1970s and 1980s by Reinhard Selten, David Kreps, Ariel Rubinstein, Avinash Dixit, Dilip Abreu and others helped fill this lacuna. Thereafter, there was no looking back. Game theory became part and parcel of modern economics and its influence stretched far and wide, to other disciplines as well as to policymaking.

5 This essay is reprinted in Kuhn and Nasar (2002). My quote is from p. 10.

6 See, for instance, Runciman and Sen (1965); Skyrms (2004); Moehler (2009); Vanderschraaf (2019, 2021).

7 For a short primer on Nash equilibrium and how it provides a foundation for much of modern economics, see Sethi and Weibull (2016).

8 The use of the Assurance Game to understand the idea of Hobbes's state of nature was proposed by Moehler (2020, chapter 2). Conventionally, the Assurance Game is treated as a two-player game. What I am arguing here is that by thinking of it as an n-player game, where $n > 2$, one gets an even better

characterization of the state of nature, where there is no coordination of action among the players.

9 See Binmore (1995, 1998).

10 See Sri Aurobindo (1910, p. 1).

CHAPTER 3

1 Nussbaum (2006), p. 173.

2 See Ressa (2022, p. xii).

3 There are philosophers and legal scholars who have argued in favour of using shame in lieu of punishments like incarceration or fines, though we need to keep in mind that punishments can vary not just in terms of pain but by the level of moral condemnation (see Kahan, 1996).

4 Ogden Nash, 'Lines fraught with naught but thought', in *Everyone But Thee and Me* (Boston: Little, Brown, 1962).

CHAPTER 4

1 I published this in 'Economic Graffiti: The Turin miracle', an op-ed for the *Indian Express*, 28 December 2017.

2 It is arguable that academic disciplines do not just describe the world but, partly, create the world that analysts then dissect, study and take away lessons from (see Mitchell, 2005, in the context of economics).

3 Kleimt (2020) takes a similar line, that modelling self-regarding opportunistic behaviour, without any room for institutionalized social norms, may be an impossibility. The 'civil economy', where people, including grassroots activists, have other-regarding preferences may not be an outlier (Becchetti and Cermelli, 2018). The general mistake which has been argued against in recent times (see Sandbu, 2020; Coyle, 2021) is to treat cultural constraints as reducible to economic interests, which are then treated as the primitives.

4 This assumption is not as innocuous as it seems at first sight. There are plausible, if paradoxical, settings where it gets violated, with important implications for moral choices (see Parfit, 1984; Basu, 2000; Voorneveld, 2010).

5 For further elaboration on this, see Basu (2000).

6 Occasionally, even when they do not change, unearthing them can help us develop novel policy interventions. The traditional view of corruption is purely in terms of fines and punishment as instruments for controlling corruption. However, once we recognize that corrupt acts have behavioural determinants, like stigma, there emerge new ways in which we can curb corrupt behaviour (see Lindbeck, Nyberg and Weibull, 1999; Lopez-Calva, 2003; Dhillon and Nicolo, 2022). Indeed, even the choice of language can activate certain moral antennae (see Capraro, Halpern and Matjiaz, 2022).

7 See Capra, Goeree, Gomez and Holt (1999); Goeree and Holt (2001), Rubinstein (2007); Brañas-Garza, Espinosa and Rey-Biel (2011); Eichberger and Kelsey (2011); Gintis (2014); Ye and Fan (2020); Conitzer and Oesterheld (2022); Izqueirdo and Izqueirdo (2023); Ramirez, Smerlak, Traulsen and Jost (2023).

8 See Basu (2018).

9 Soon after I wrote these lines, a controversy erupted about the World Bank having given in to pressure from China and Saudi Arabia and tampered with its Ease of Doing Business rankings (after I left the World Bank, I hasten to add). There was an independent inquiry by the law firm WilmerHale, and the charges were serious enough that the World Bank's board decided in September 2021 to stop its annual Doing Business exercise.

10 See Kamtekar (2012, p. 170).

11 See Basu (2000, appendix A6). This is a debate that goes far back to the ancient Greeks, examining the 'compatibilism' between determinism and responsibility (see Bobzien, 1998; Brennan, 2001).

12 This is the idea behind the concept of 'conferred morality' that I describe in Basu (2022a).

13 See Bobzien (1998). As Brennan (2001, p. 263) puts it, Chrysippus constructed a 'doctrine of determinism' and 'a doctrine of moral responsibility' which do not negate each other.

14 'Little Boxes', words and music by Malvina Reynolds; copyright 1962 Schroder Music Company, renewed 1990.

15 Interestingly, controlled experiments show that people have a natural propensity to be guided by efficiency considerations when deciding whom to punish and how (see Handfield, Thrasher, Corcoran and Nichols, 2021).

16 Indeed, it is arguable that a nurse can in fact do her job better if she can 'not be emotionally distressed by the patient's pain' (Heydari Fard, 2018, p. 81). While this has other complications, since the *sight* of her pain can be comforting to others, it illustrates how curbing a painful emotion but nevertheless taking the empathetic action that the emotion would have prompted is the direction to go.

CHAPTER 5

1 In an earlier work, Basu (2000, chapter 4), I discussed at length why we do not see trade and exchange in the world of rats, why the presumptions of mainstream economics may not be right, and how without a modicum of trust and honesty markets are likely to break down (see also Arrow, 1978).

2 See Havel (1986).

3 See Akerlof (1976). The broader idea of 'triadic forces' that arise from this in different contexts has been investigated in several papers: see Hatlebakk (2002); Villanger (2006); Li (2023).

4 There is a literature, with allegorical constructions, and spanning different disciplines, which has highlighted this problem. See Chocker and Halpern (2004); Copp (2006); Chiao (2014); Chant (2015); Friedenberg and Halpern (2019); Hyska (2021).

5 Dave King, 'The Immoral Democrat Party "List of Shame"', *Conservative Daily News*, 6 November 2015.

6 Nancy Altman and Linda Benesch, 'The Deeply Immoral Values of Today's Republican Leaders', *Huffington Post*, 11 December 2017.

7 See, for instance, Feinberg (1968); Bernheim and Whinston (1986); Marino (2001); Sartorio (2004); Tännsjö (2007); Petersson (2008); Hakli, Miller and Tuomela (2010); List and Pettit (2011); Bjornsson (2014); Hess (2014); Arruda (2017); Dughera and Marciano (2020).

8 In the language of Marino (2001), individuals can be 'blameless' but belong to a group that has *group* moral responsibility. Further, there are situations where, by holding a person *legally* responsible for some outcome, we may be able to change the person's behaviour for the better. Tort liability is often concerned with this. However, even without invoking the law, at times it is worthwhile to blame a person for an outcome even when this does not reflect her ill will or culpable ignorance, because the human aversion to blame can change behaviour and help bring about a better outcome (see Hankins, 2016).

9 See, for instance, Frankfurt (1969); Haji and McKenna (2004); Pereboom (2017).

10 This is closely related to the idea of choosing an action for the right reason. The reason for acting in a certain way must be the reason that makes this action morally right (see Markovits, 2010).

11 See Basu (2022a, 2022b); Heydari Fard (2022).

12 The remainder of this section draws on my paper, Basu (2022b).

13 While this specific game is illustrative, it is arguable that this is a real problem which has implications for how we shape policy in the real world. It should be possible to do some interesting laboratory tests with Greta's Dilemma to see if the predicted moral setback is likely in reality. Such tests should take the

form of making players play the basic game in Table 1, without letting them know of the effect their play has on the bystander. Thereafter, tell Player 1 about the bystander and maybe even do some priming on moral responsibility, and then make them play the same game, to see if Player 1 flips from choosing B to A. There has been work done concerning the manipulation of beliefs and focal points in laboratory settings (see Dasgupta and Radoniqi, 2021), which can suggest ways of imputing morals and values into actions.

14 In the context of game theory and morality, related ideas are discussed in Bacharach (1999); Braham and Holler (2009); Braham and van Hees (2012).

15 A related problem has been discussed in coalition-formation literature (see Aumann and Myerson, 1988; Genicot and Ray, 2003; Ray and Vohra, 2015). But the choice of *morals* raises conceptually distinct matters.

16 See Clark and Chalmers (1998); Fioretti and Policarpi (2020).

17 See Putnam (2005, p. 24).

18 Morals can be of many kinds. There is a growing literature in economics on this. See Bowles and Gintis, 1998; Alger and Weibull, 2013; Bowles, 2016; Sen, 2018.

19 One way out is if everybody has the morality and chooses from behind a veil of ignorance, so that you do not know who you will be in the game. Such a game would be described as a unanimity game (Basu, 2010). In that case, the problem gets resolved in the sense that the outcome that gives the highest payoff is always part of an equilibrium. Bill Gates once wrote (*The Economist*, 16 February 2019, p. 62, column 1) about the importance of going beyond self-interest and how we ought to have moral causes that we stand up for: ‘If you want to improve the world, you need something to be mad about.’ What we just showed is that if everyone takes this to heart and is mad about a larger cause, the world may end up a worse place. The only way out is for everyone to be mad in the same way.

CHAPTER 6

1 Quoted from Voltaire's 1771 essay on 'Rights'.

2 See Vanderschraaf (2006); Moehler (2009).

3 There are also statistical studies over sweeps of history to puzzle over and understand what makes civil wars happen, and what foils them. Fearon and Laitin (2014) analyse the conundrum of post-war Japan, where rebellion never broke out despite periods of intense grievance. They analyse this in terms of the strength of the state. This is important but, as the analysis below shows, the *mechanism* of control matters greatly and many a strong state fails on that front.

4 See Schelling (1960); Mehta, Starmer and Sugden (1994); Sugden (1995).

5 As Boumlik and Schwartz (2016) noted, 'Yahyaoui's influence within the collective movement contributed to regime change through her advocacy from her exile in France via social networks.'

6 We can make this more realistic by saying that unless the probability of being jailed for protesting is higher than some number, p (>0), a person will protest. But no harm is done to the logic I am about to demonstrate by making this more extreme assumption, namely, that unless one is certain one will be jailed for protesting, one will protest. Any probability of being jailed that is less than 1 makes people want to go out and protest. For simplicity, I shall also assume that each person treats the success of the revolution as independent of one's own action. In a large population this is not unreasonable, and it keeps our analysis simple. So the satisfaction from protesting is the pure joy of participating in this important task of helping the nation. This keeps us out of the public-goods problem that concerned Tullock (1971) and, anyway, it is not germane to what I am arguing here. It should be emphasized that my assumption is not unrealistic. People do get satisfaction from voting even when they know their individual vote is unlikely to

matter; people do feel good not throwing plastic into the river even though one piece of plastic is unlikely to do any harm.

7 See O'Connor (1948); Quine (1953); Scriven (1951); Levy (2009). Backward induction and iterated reasoning are widely used in game theory, such as the Centipede game and the Traveller's Dilemma, and have been the subject of philosophical scrutiny (Rosenthal, 1981; Bicchieri, 1989; Basu, 1994; Rubinstein, 2006; Arad and Rubinstein, 2012; Halpern and Pass, 2012; Alaoui and Penta, 2016; Asheim and Brunnschweiler, 2023).

8 The pervasive use of these kinds of knowledge hierarchies, from various academic disciplines to British whodunits and real-life situations, like eating pau bhaji on the beach, are discussed in Sarangi (2020).

9 This argument is more pervasive in life than may appear at first sight. In keeping with a tradition in economics, I am using this analysis of dictators in the context only of the state and political leadership, but some of this can happen within corporations and other organizations, even though mainstream economics lacks the vocabulary for it (see Anderson, 2017).

10 If this problem were applied to a nation with a countably infinite population, common knowledge would be necessary. Anything short of infinite layers of knowledge would not work.

11 See Muldoon (2019); Moehler (2020).

12 In much of the literature on collective action, in economics and also philosophy, it is assumed that individuals are self-serving. In reality, people are often other-regarding and benevolent. This can enable us to formalize the idea of activism and give insights into constitutional design (see Mishra and Anant, 2006; Singh, 2006; Becchetti and Cermelli, 2018).

13 In Basu (2010) I try to show that the invisible hand can take different forms, from Adam Smith's benevolent invisible hand of the market to Franz Kafka's invisible hand that can create political oppression without giving us the comfort of being able to point a finger at the perpetrator.

CHAPTER 7

1 'September 1, 1939', W. H. Auden, first published in *The New Republic,* 18 October 1939.

2 Though this precise sentence was not a part of the Manifesto but Marx's later essay 'Critique of the Gotha Programme', published in 1875.

3 See Aquinas (1265–74, II–II, 66.7).

4 Basu (2022c, 2023) analyses the power of political leaders and why there may be an innate dynamic which leads to authoritarian leaders morphing into oppressive dictators. This provides grist to the argument that term limits for political leaders should be a requirement in all nations.

5 See Basu (2021).

6 The subject of norms exercising restraint on behaviour has a long literature, but only now is it beginning to encroach on mainstream economic theory (see Sunstein, 1996; Posner, 1998; Schlicht, 1998; Basu, 2000; Richter and Rubinstein, 2020).

7 As Acemoglu and Robinson (2019, p. 467) observe in the same section, we 'need the state to play a role in redistribution, creating a social safety net and regulating the increasingly complex economy . . .'

8 See Stiglitz (2012); Piketty (2014, 2020); Atkinson (2015); Bourguignon (2015).

9 See Sherman (2019).

10 This is based on the assumption of an income of 5 per cent of the wealth per annum. For statistical analysis of global inequality, see Bourguignon (2015); Milanovic (2018).

11 The number is not official data but highly plausible. Consider individuals with wealth of $100 billion. We know there is a small number in this category. If they earn an annual income which is 5 per cent of this wealth, this too is plausible. They would not be so rich if they were doing much worse than this. That, in turn, amounts to an income of $13,690,000 a day.

12 See, for instance, Banerjee and Newman (1993); Galor and Zeira (1993); Emerson and Souza (2003).

13 See Coyle (2021); Basu, Caspi and Hockett (2021).

14 See Schaefer and Singh (2022).

15 See Naidu, Posner and Weyl (2018).

16 See Basu, Caspi and Hockett (2021).

17 Related ideas, whereby poverty is tackled while simultaneously attending to inequality, have been explored in the literature, even though this particular tax, which has a stark simplicity and is rooted in behavioural economics, is arguably new (see Sen, 1973; Bourguignon and Fields, 1990; Jayaraj and Subramanian, 1996; Basu, 2006; Subramanian, 2006). What I am about to discuss also has some roots in the philosophy of prioritarianism (see, for instance, Parfit, 2000; Adler, 2022) and limitarianism (see Robeyns, 2019).

18 See Hughes (2010, p. 5).

19 See Navia (2007, p. 105).

20 Not only are rankings unaffected, the rankings of income gaps will be unaffected. If the gap between persons I's and J's incomes is greater than the gap between persons K's and L's incomes pre-tax, it will be greater also after tax. And, in fact, gaps between gaps will also be unchanged and all higher-order gaps, in the spirit of cardinal utility (see Basu, 1983).

21 See Ferguson (2013); Vanderschraaf (2019).

22 This has been referred to as a 'second-order collective action problem' (Ferguson, 2020). See also Khan (2018) for the role of political coordination as a substitute for violence and conflict.

23 The core idea of this is developed in Basu and Weibull (1991). This is related to the idea of having limits on the behaviour of the state which are self-enforcing (see Weingast, 1997). It is worth pointing out that in reality people are often willing to go beyond strictly self-regarding behaviour for the collective good, such as for managing the village commons and halting

resource degradation (see Baland and Platteau, 1996). Surveys and experimental studies show that people have the capacity for 'conditional cooperation', the capacity to make personal sacrifice if others are willing to do the same (see Rustagi, Engel and Kosfeld, 2010).

24 See Rawls (1997); Weithman (2005); Thrasher and Vallier (2013); Sayre-McCord (2017).

25 There are controlled studies done at a micro-level, which show how communication and promises influence the way people actually behave (Charness and Dufwenberg, 2006; see also Bahel, Ball and Sarangi, 2022). It will be interesting to try to take this idea to the level of public discourse and public deliberation, à la Weithman (2005), to see if these commitments can be made even stronger.

26 See Beckerman (2022).

27 Letter of 20 September 1953 (see Khosla, 2014, p. 67).

28 See Bowles and Gintis (1998, p. 4).

29 See Sen (2006) for an important analysis of our multiple identities and the risks and hopes that arise from them.

30 On nationalism, Nehru had a visionary ally in the poet Rabindranath Tagore, who wrote passionately reminding people of their larger human identity that went beyond nation. It is a fitting tribute to Tagore that he may be the only person in history whose songs have been adopted for their national anthem by more than one country: India and Bangladesh.

31 See Niemi and Young (2013). There are studies that use eye-tracking to study differences in attitudes to in-groups and out-groups among people of different degrees of pro-sociality. The connections are complex but it is interesting that there are marked differences (see Rahal, Fiedler and de Dreu, 2020).

References

Acemoglu, D. and Robinson, J. (2019), *The Narrow Corridor: States, Society, and the Fate of Liberty*, New York: Penguin Press.

Adler, M. (2022), 'Theory of Prioritarianism', in M. Adler and O. F. Norheim (eds.), *Prioritarianism in Practice*, Cambridge: Cambridge University Press.

Akerlof, G. (1970), 'The Market for "Lemons": Quality Uncertainty and the Market Mechanism', *Quarterly Journal of Economics*, 84: 488–500.

Akerlof, G. (1976), 'The Economics of Caste and of the Rat Race and Other Woeful Tales', *Quarterly Journal of Economics*, 90, 4: 599–617.

Akerlof, R. (2016), 'Anger and Enforcement', *Journal of Economic Behavior and Organization*, 126: 110–24.

Alaoui, L. and Penta, A. (2016), 'Endogenous Depth of Reasoning', *Review of Economic Reasoning*, 83, 4: 1297–333.

Alger, I. and Weibull, J. (2013), '*Homo Moralis* – Preference Evolution under Incomplete Information and Assortative Matching', *Econometrica*, 81: 2269–302.

Anderson, E. (2017), *Private Government: How Employers Rule Our Lives (and Why We Don't Talk about It)*, Princeton: Princeton University Press.

Aquinas, T. (1265–74), *Summa Theologica*, Allen, TX: Christian Classics (1911).

Arad, A. and Rubinstein, A. (2012), 'The 11-20 Money Request

Game: A Level-k Reasoning Study', *American Economic Review*, 102, 7: 3561–73.

Arrow, K. J. (1978), 'A Cautious Case for Socialism,' *Dissent*, September: 472–82.

Arrow, K. J. and Debreu, G. (1954), 'Existence of an Equilibrium for a Competitive Economy', *Econometrica*, 22: 265–90.

Arruda, C. (2017), 'How I Learned to Worry about the Spaghetti Western: Collective Responsibility and Collective Agency', *Analysis*, 77, 2: 249–59.

Asheim, G. B. and Brunnschweiler, T. (2023), 'Epistemic Foundation of the Backward Induction Paradox', *Games and Economic Behavior*, forthcoming.

Atkinson, A. B. (2015), *Inequality: What Can Be Done*, Cambridge, MA: Harvard University Press.

Aumann, R. J. (1976), 'Agreeing to Disagree', *Annals of Statistics*, 4: 1236–9.

Aumann, R. J. and Myerson, R. (1988), 'Endogenous Formation of Links between Players and of Coalitions: An Application of the Shapley Value', in A. Roth (ed.), *The Shapley Value: Essays in Honor of Lloyd S. Shapley*, Cambridge: Cambridge University Press.

Bacharach, M. (1999), 'Interactive Team Reasoning: A Contribution to the Theory of Co-operation', *Research in Economics*, 53, 2: 117–47.

Bahel, E., Ball, S. and Sarangi, S. (2022), 'Communication and Cooperation in Prisoner's Dilemma Games', *Games and Economic Behavior*, 133: 126–37.

Baland, J. M. and Platteau, J. P. (1996), *Halting Degradation of Natural Resources: Is There a Role for Rural Communities?*, Oxford: Oxford University Press.

Banerjee, A. and Newman, A. (1993), 'Occupational Choice and the Process of Development', *Journal of Political Economy*, 101, 2: 274–98.

Basu, K. (1983), 'Cardinal Utility, Utilitarianism and a Class of Invariance Axioms in Welfare Analysis', *Journal of Mathematical Economics*, 12, 3: 193–206.

Basu, K. (1994), 'The Traveler's Dilemma: Paradoxes of Rationality in Game Theory', *American Economic Review, Papers and Proceedings*, 71: 391–5.

Basu, K. (2000), *Prelude to Political Economy: A Study of the Social and Political Foundations of Economics*, Oxford and New York: Oxford University Press.

Basu, K. (2006), 'Globalization, Poverty and Inequality: What Is the Relationship? What Can be Done?', *World Development*, 34, 8: 1361–73.

Basu, K. (2010), *Beyond the Invisible Hand: Groundwork for a New Economics*, Princeton: Princeton University Press.

Basu, K. (2014), 'Randomization, Causality and the Role of Reasoned Intuition', *Oxford Development Studies*, 42, 4: 455–72.

Basu, K. (2017), 'A New and Rather Long Proof of the Pythagorean Theorem by Way of a Proposition on Isosceles Triangles', *The College Mathematics Journal*, 47, 5: 356–60.

Basu, K. (2018), *The Republic of Beliefs: A New Approach to Law and Economics*, Princeton: Princeton University Press.

Basu, K. (2021), 'The Ground beneath Our Feet', *Oxford Review of Economic Policy*, 37, 4: 783–93.

Basu, K. (2022a), 'The Samaritan's Curse: Moral Individuals and Immoral Groups', *Economics & Philosophy*, 38, 1: 132–51.

Basu, K. (2022b), 'Conventions, Morals, and Strategy: Greta's Dilemma and the Incarceration Game', *Synthese*, 200, 1: 1–19.

Basu, K. (2022c), 'Why Have Leaders at All? Hume and Hobbes with a Dash of Nash', *Homo Oeconomicus*, 39.

Basu, K. (2023), 'The Morphing of Dictators: Why Dictators get Worse over Time', *Oxford Open: Economics*, 2.

Basu, K., Caspi, A. and Hockett, R. (2021), 'Markets and Regulation in the Age of Big Tech', *Capitalism and Society*, 15, 1.

Basu, K. and Weibull, J. (1991), 'Strategy Subsets Closed under Rational Behavior', *Economics Letters*, 36, 2: 141–6.

Becchetti, L. and Cermelli, A. M. (2018), 'Civil Economics: Definition and Strategies for Sustainable Well-Being', *International Review of Economics*, 65: 1–29.

Beckerman, G. (2022), *The Quiet Before: On the Unexpected Origins of Radical Ideas*, New York: Crown.

Bernheim, B. D. and Whinston, M. D. (1986), 'Common Agency', *Econometrica*, 54, 4: 923–42.

Bertrand, J. (1883), 'Review of "Théorie mathématique de la richesse sociale" and "Recherche sur les principes mathématiques de la théorie des richesses"', *Journal des Savants*, 67: 499–508.

Bicchieri, C. (1989), 'Self-refuting theories of Strategic Interaction: A Paradox of Common Knowledge', *Erkenntnis*, 30: 69–85.

Binmore, K. (1995), 'The Game of Life: Comment', *Journal of Institutional and Theoretical Economics*, 151: 132–6.

Binmore, K. (1998), *Just Playing: Game Theory and the Social Contract II*, Cambridge, MA: MIT Press.

Bjornsson, G. (2014), 'Essentially Shared Obligations', *Midwest Studies in Philosophy*, 38: 103–20.

Bobzien, S. (1998), *Determinism and Freedom in Stoic Philosophy*, Oxford: Oxford University Press.

Boumlik, H. and Schwartz, J. (2016), 'Conscientization and Third Space: A Case Study of Tunisian Activism', *Adult Education Quarterly*, 66, 4: 319–35.

Bourguignon, F. (2015), *The Globalization of Inequality*, Princeton: Princeton University Press.

Bourguignon, F. and Fields, G. (1990), 'Poverty Measures and Anti-Poverty Policy', *Recherches Économiques de Louvain*, 56, 3–4: 409–27.

Bowles, S. (2016), *The Moral Economy: Why Good Incentives are No Substitute for Good Citizens*, New Haven: Yale University Press.

Bowles, S. and Gintis, H. (1998), 'The Moral Economy of Communities: Structured Populations and the Evolution of Pro-Social Norms', *Evolution and Human Behavior*, 19, 1: 3–25.

Braham, M. and Holler, M. J. (2009), 'Distributing Causal Responsibility in Collectivities', in R. Gekker and T. Boylan (eds.), *Economics, Rational Choice and Normative Philosophy*, New York: Routledge.

Braham, M. and van Hees, M. (2012), 'An Anatomy of Moral Responsibility', *Mind*, 121, 483: 601–34.

Brañas-Garza, P., Espinosa, M. P. and Rey-Biel, P. (2011), 'Travelers' Types', *Journal of Economic Behavior & Organization*, 78, 1–2: 25–36.

Brennan, G. and Sayre-McCord, G. (2018), 'On "Cooperation"', *Analyse & Kritik*, 40, 1: 107–30.

Brennan, T. (2001), 'Fate and Free Will in Stoicism', in D. Sedley (ed.), *Oxford Studies in Ancient Philosophy*, Oxford: Oxford University Press.

Capra, M., Goeree, J., Gomez, R. and Holt, C. A. (1999), 'Anomalous Behavior in a Traveler's Dilemma?', *American Economic Review*, 89, 3: 678–90.

Capraro, V., Halpern, J. and Matjiaz, P. (2022), 'From Outcome-Based to Language-Based Preferences', *Journal of Economic Literature*, forthcoming.

Cartwright, N. (2010), 'What Are Randomized Trials Good For?', *Philosophical Studies*, 147: 59–70.

Chant, S. R. (2015), 'Collective Responsibility in a Hollywood Standoff', *Thought: A Journal of Philosophy*, 4, 2: 83–92.

Charness, G. and Dufwenberg, M. (2006), 'Promises and Partnership', *Econometrica*, 74, 6: 1579–601.

Chiao, V. (2014), 'List and Pettit on Group Agency and Group Responsibility', *University of Toronto Law Journal*, 64, 5: 753–70.

Chocker, H. and Halpern, J. Y. (2004), 'Responsibility and Blame: A Structural-Model Approach', *Journal of Artificial Intelligence Research*, 22: 93–115.

Christiansen, M. and Chater, N. (2022), *The Language Game: How Improvisation Created Language and Changed the World*, New York: Basic Books.

Clark, A. and Chalmers, D. (1998), 'The Extended Mind', *Analysis*, 58, 1: 7–19.

Conitzer, V. and Oesterheld, C. (2022), 'Foundations of Cooperative AI', mimeo: Carnegie Mellon University.

Copp, D. (2006), 'On the Agency of Certain Collective Entities: An Argument from "Normative Autonomy"', *Midwest Studies in Philosophy*, 30, 1: 194–221.

Cortina, A. (2022), *Aporophobia: Why We Reject the Poor Instead of Helping Them*, Princeton: Princeton University Press. [Spanish orig., *Aporofobia, el rechazo al pobre: Un desafío para la democracia* (2017)].

Cournot, A. A. (1838), *Recherches sur les principes mathématiques de la théorie des richesses*, Paris: Hachette. [Eng. tr.: *Researches into the Mathematical Principles of the Theory of Wealth*, tr. Nathaniel T. Bacon (New York: Macmillan, 1927)].

Coyle, D. (2021), *Cogs and Monsters: What Economics Is and What It Should Be*, Princeton: Princeton University Press.

Dasgupta, U. and Radoniqi, F. (2021), 'Republic of Beliefs: An Experimental Investigation', IZA Discussion Paper, No. 14130.

Debreu, G. (1959), *Theory of Value: An Axiomatic Analysis of Economic Equilibrium*, New Haven: Yale University Press.

Dhillon, A. and Nicolo, A. (2022), 'Moral Costs of Corruption', in K. Basu and A. Mishra (eds.), *Law and Economic Development: Behavioral and Moral Foundations of a Changing World*, New York: Palgrave Macmillan.

Dixit, A. (2004), *Lawlessness and Economics: Alternative Modes of Governance*, Princeton: Princeton University Press.

Dughera, S. and Marciano, A. (2020), 'Self-Governance, Non-reciprocal Altruism and Social Dilemmas', mimeo: University of Paris Nanterre.

Eichberger, J. and Kelsey, D. (2011), 'Are the Treasures of Game Theory Ambiguous?', *Economic Theory*, 48, 2/3: 313–39.

Elster, J. (1989), *The Cement of Society: A Study of Social Order*, Cambridge: Cambridge University Press.

Emerson, P. and Souza, A. P. (2003), 'Is There a Child Labor Trap? Intergenerational Persistence of Child Labor in Brazil', *Economic Development and Cultural Change*, 51, 2: 375–98.

Fearon, J. and Laitin, D. (2014), 'Civil War Non-Onsets: The Case of Japan', *Medeniyet Arastirmalari Dergisi*, 1, 1: 71–94.

Feinberg, J. (1968), 'Collective Responsibility', *Journal of Philosophy*, 65, 21: 674–88.

Feld, S. (1991), 'Why Your Friends Have More Friends Than You Do', *American Journal of Sociology*, 96, 6: 1464–77.

Ferguson, W. (2013), *Collective Action and Exchange: A Game-theoretic Approach to Contemporary Political Economy*, Stanford: Stanford University Press.

Ferguson, W. (2020), *The Political Economy of Collective Action, Inequality, and Development*, Stanford: Stanford University Press.

Fioretti, G. and Policarpi, A. (2020), 'The Less Intelligent the Elements, the More Intelligent the Whole. Or Possibly Not?', mimeo: University of Bologna.

Frank, R. (2020), *Under the Influence: Putting Peer Pressure to Work*, Princeton: Princeton University Press.

Frankfurt, H. G. (1969), 'Alternative Possibilities and Moral Responsibility', *Journal of Philosophy*, 66, 23: 829–39.

Friedenberg, M. and Halpern, J. (2019), 'Blameworthiness in Multi-Agent Settings', *Association for the Advancement of Artificial Intelligence*, 33, 1: 525–32.

Galor, O. and Zeira, J. (1993), 'Income Distribution and Macroeconomics', *Review of Economic Studies*, 60, 1: 35–52.

Genicot, G. and Ray, D. (2003), 'Group-Formation in Risk-Sharing Arrangements', *Review of Economic Studies*, 70, 1: 87–113.

Gintis, H. (2014), *The Bounds of Reason: Game Theory and the Unification of the Behavioral Sciences*, Princeton: Princeton University Press.

Goeree, J. and Holt, C. (2001), 'Ten Little Treasures of Game Theory and Ten Intuitive Contradictions', *American Economic Review*, 91, 5: 1402–22.

Goffman, E. (1963), *Stigma: Notes on the Management of Spoiled Identity*, New York: Simon and Schuster.

Haji, I. and McKenna, M. (2004), 'Dialectical Delicacies in the Debate about Freedom and Alternative Possibilities', *Journal of Philosophy*, 101, 6: 299–314.

Hakli, R., Miller, K. and Tuomela, R. (2010), 'Two-kinds of We-reasoning', *Economics & Philosophy*, 26, 3: 291–320.

Halpern, J. and Pass, R. (2012), 'Iterated Regret Minimization: A New Solution Concept', *Games and Economic Behavior*, 74: 153–8.

Handfield, T., Thrasher, J., Corcoran, A. and Nichols, S. (2021), 'Asymmetry and Symmetry of Acts and Omissions in Punishment, Norms and Judged Causality', *Judgement and Decision Making*, 16, 4: 796–822.

Hankins, K. (2016), 'Adam Smith's Intriguing Solution to the Problem of Moral Luck', *Ethics*, 126, 3: 711–46.

Hatlebakk, M. (2002), 'A New and Robust Model of Subgame Perfect Equilibrium in a Model of Triadic Power Relations', *Journal of Development Economics*, 68, 1: 225–32.

Havel, V. (1986), 'The Power of the Powerless', in J. Vladislav (ed.), *Living in Truth*, London: Faber & Faber.

Hayek, F. A. (1944), *The Road to Serfdom*, London: Routledge.

Hess, K. (2014), 'Because They Can: The Basis for Moral Obligations for (Certain) Collectives', *Midwest Studies in Philosophy*, 38: 203–21.

Heydari Fard, S. (2018), 'Decision-theoretic Consequentialism and the Desire-Luck Problem', *Journal of Cognition and Neuroethics*, 5(1): 77–84.

Heydari Fard, S. (2022), 'Strategic Injustice, Dynamic Network Formation and Social Movements', *Synthese*, 200(5).

Hobbes, T. (1651), *Leviathan*, ed. R. Tuck, Cambridge: Cambridge University Press (1991).

Hoeft, L. (2019), 'The Force of Norms? The Internal Point of View in Light of Experimental Economics', *Ratio Juris*, 32, 3: 339–62.

Hughes, B. (2010), *The Hemlock Cup: Socrates, Athens and the Search for the Good Life*, London: Jonathan Cape.

Hume, D. (1740), *A Treatise of Human Nature*, eds D. F. Norton and M. J. Norton, Oxford: Oxford University Press (2000).

Hyska, M. (2021), 'Propaganda, Irrationality and Group Agency', in M. Hannon and J. de Ridder (eds), *The Routledge Handbook of Political Epistemology*, London: Routledge.

Izquierdo, S. and Izqueirdo, L. (2023), 'Strategy Sets Closed under Payoff Sampling', *Games and Economic Behavior*, 138: 126–42.

Jayaraj, D. and Subramanian, S. (1996), 'Poverty-Eradication through Redistributive Taxation: Some Elementary Considerations', *Review of Development and Change*, 1: 73–84.

Kahan, D. (1996), 'What do Alternative Sanctions Mean?', *University of Chicago Law Review*, 63(2): 591–653.

Kamtekar, R. (2012), 'Speaking with the Same Voice as Reason: Personification in Plato's Psychology', in R. Barney, T. Brennan and C. Brittain (eds), *Plato and the Divided Self*, Cambridge: Cambridge University Press.

Khan, M. (2018), 'Political Settlements and the Analysis of Institutions', *African Affairs*, 117, 469: 636–55.

Khosla, M. (ed.) (2014), *Letters for a Nation: From Jawaharlal Nehru to His Chief Ministers, 1947–1963*, New Delhi: Penguin.

Kleimt, H. (2020), 'Economic and Sociological Accounts of Social Norms', *Analyse & Kritik*, 42, 1: 41–95.

Kuhn, H. and Nasar, S. (2002), *The Essential John Nash*, Princeton: Princeton University Press.

Larkin, P. (1982), 'The Art of Poetry', interview by Robert Phillips, *Paris Review*, 84.

Levy, K. (2009), 'The Solution to the Surprise Exam Paradox', *Southern Journal of Philosophy*, 47, 2: 131–58.

Lewis, D. (1969), *Convention: A Philosophical Study*, Cambridge, MA: Harvard University Press.

Li, C. (2023), *Essays on Network Supervision, On-line Price Signalling, and E-Commerce Effects on Offline Stores*, PhD thesis, Cornell University.

Lindbeck, A., Nyberg, S. and Weibull, J. (1999), 'Social Norms and Economic Incentives in the Welfare State', *Quarterly Journal of Economics*, 114, 1: 1–35.

List, C. and Pettit, P. (2011), *Group Agency: The Possibility, Design, and the Status of Corporate Agents*, Oxford: Oxford University Press.

Lopez-Calva, L. F. (2003), 'Social Norms, Coordination and Policy Issues in the Fight Against Child Labor', in K. Basu, M. Horn, L. Roman and J. Shapiro (eds), *International Labour Standards*, Oxford: Blackwell.

Mailath, G., Morris, S. and Postlewaite, A. (2017), 'Laws and Authority', *Research in Economics*, 71: 32–42.

Marino, P. (2001), 'Moral Dilemma, Collective Responsibility and Moral Progress', *Philosophical Studies*, 104, 2: 203–25.

Markovits, J. (2010), 'Acting for the Right Reasons', *Philosophical Review*, 119, 2: 201–42.

May, E. R. and Zelikow, P. D. (1997), *The Kennedy Tapes*, Cambridge, MA: Harvard University Press.

Mehta, J., Starmer, C. F. and Sugden, R. (1994), 'Focal points in pure coordination games: An experimental investigation', *Theory and Decision*, 36: 163–85.

Milanovic, B. (2018), *Global Inequality: A New Approach for the Age of Globalization*, Cambridge, MA: Harvard University Press.

Miller, A. D. and Perry, R. (2012), 'The Reasonable Person', *New York University Law Review*, 87: 323–92.

Mishra, A. and Anant, T. C. A. (2006), 'Activism, Separation of Powers and Development', *Journal of Development Economics*, 81, 2: 457–77.

Mitchell, T. (2005), 'The Work of Economics: How a Discipline Makes Its World', *European Journal of Sociology*, 46, 2: 297–320.

Moehler, M. (2009), 'Why Hobbes' State of Nature Is Best Modeled as an Assurance Game', *Utilitas*, 21, 3: 297–326.

Moehler, M. (2020), *Contractarianism*, Cambridge: Cambridge University Press.

Muldoon, R. (2019), *Social Contract Theory for a Diverse World: Beyond Tolerance*, New York: Routledge.

Myerson, R. (1991), *Game Theory: Analysis of Conflict*, Cambridge, MA: Harvard University Press.

Myerson, R. (2004), 'Justice, Institutions and Multiple Equilibria', *Chicago Journal of International Law*, 5: 91–107.

Naidu, S., Posner, E. and Weyl, G. (2018), 'Antitrust Remedies for Labor Market Power', *Harvard Law Review*, 132, 2: 536–601.

Nash, J. (1950), 'Equilibrium Points in n-Person Games', *PNAS*, 36, 1: 48–9.

Navia, L. E. (2007), *Socrates: A Life Examined*, Amherst: Prometheus Books.

Niemi, L. and Young, L. (2013), 'Caring across Boundaries versus Keeping Boundaries Intact: Links between Moral and Interpersonal Orientations', *PloS ONE*, 8, 12: 1–12.

Nussbaum, M. (2006), *Hiding from Humanity: Disgust, Shame and the Law*, Princeton: Princeton University Press.

O'Connor, D. J. (1948), 'Pragmatic Paradoxes', *Mind*, 57, 227: 358–9.

Parfit, D. (1984), *Reasons and Persons*, Oxford: Clarendon Press.

Parfit, D. (2000), 'Equality or Priority?', in M. Clayton and A. Williams (eds), *The Ideal of Equality*, New York: Palgrave Macmillan.

Pereboom, D. (2017), 'Responsibility, Regret and Protest', in D. Shoemaker (ed.), *Oxford Studies in Agency and Responsibility*, Oxford: Oxford University Press.

Petersson, B. (2008), 'Collective Omissions and Responsibilities', *Philosophical Papers*, 37, 2: 243–61.

Piketty, T. (2014), *Capital in the 21st Century*, Cambridge, MA: Harvard University Press.

Piketty, T. (2020), *Capital and Ideology*, Cambridge, MA: Harvard University Press.

Platteau, J. P. (2000), *Institutions, Social Norms, and Economic Development*, Amsterdam: Harwood Academic Publishers.

Posner, E. (1998), 'Law, Economics, and Inefficient Norms', *University of Pennsylvania Law Review*, 144: 1697–744.

Putnam, H. (2005), *Ethics without Ontology*, Harvard University Press, Cambridge, MA.

Quine, W. V. (1953), 'On a So-called Paradox', *Mind*, 62, 245: 65–7.

Rahal, R. M., Fiedler, S. and De Dreu, C. K. W. (2020), 'Prosocial Preferences Condition Decision Effort and Ingroup biased Generosity in Intergroup Decision-Making', *Scientific Reports*, 10, 1: 10132.

Ramirez. M., Smerlak, M., Traulsen, A. and Jost, J. (2023), 'Diversity Enables the Jump towards Cooperation for the Traveler's Dilemma', *Scientific Reports*, 13, 1441.

Rasmussen, D. (2017), *The Infidel and the Professor: David Hume, Adam Smith, and the Friendship that Shaped Modern Thought*, Princeton: Princeton University Press.

Ratner, B. (2009), 'Pythagoras: Everyone knows his famous theorem but not who discovered it 1000 years before him', *Journal of Targeting, Measurement and Analysis for Marketing*, 17: 229–42.

Rawls, J. (1971), *A Theory of Justice*, Cambridge, MA: Harvard University Press.

Rawls, J. (1997), 'The Idea of Public Reasoning Revisited', *University of Chicago Law Review*, 64, 3: 765–807.

Ray, D. and Vohra, R. (2015), 'Coalition Formation', in R. J. Aumann and S. Hart (eds), *Handbook of Game Theory with Economic Applications*, Amsterdam: Elsevier.

Ressa, M. (2022), *How to Stand up to a Dictator: The Fight for Our Future*, London: WH Allen.

Richter, M. and Rubinstein, A. (2020), 'The Permissible and the Forbidden', *Journal of Economic Theory*, 188: 105–42.

Robeyns, I. (2019), 'What if Anything is Wrong with Extreme Wealth?', *Journal of Human Development and Capabilities*, 20(3): 251–66.

Rosenthal, R. W. (1981), 'Games of Perfect Information, Predatory Pricing and the Chain Store Paradox', *Journal of Economic Theory*, 25: 92–100.

Rubinstein, A. (2006), 'Dilemmas of an Economic Theorist', *Econometrica*, 74: 865–83.

Rubinstein, A. (2007), 'Instinctive and Cognitive Reasoning: A Study of Response Times', *Economic Journal*, 117: 1243–59.

Runciman, W. and Sen, A. (1965), 'Games, Justice and the General Will', *Mind*, 74, 296: 554–62.

Russell, B. (1903), *Principles of Mathematics*, Cambridge: Cambridge University Press.

Russell, B. (1912), *The Problems of Philosophy*, London: Williams and Norgate.

Russell, B. (1946), *History of Western Philosophy*, London: Allen and Unwin.

Rustagi, D., Engel, S. and Kosfeld, M. (2010), 'Conditional

Cooperation and Costly Monitoring Explain Success in Forrest Commons Management', *Science*, 330, 6006: 961–5.

Samuelson, L. and Stacchetti, E. (2017), 'Even Up: Maintaining Relationships', *Journal of Economic Theory*, 169: 170–217.

Sandbu, M. (2020), *The Economics of Belonging*, Princeton: Princeton University Press.

Sarangi, S. (2020), *The Economics of Small Things*, New Delhi: Penguin.

Sartorio, C. (2004), 'How to Be Responsible for Something Without Causing It', *Philosophical Papers*, 18: 315–36.

Sayre-McCord, G. (2017), 'Hume's Theory of Public Reason', in P. N. Turner and G. Gaus (eds), *Public Reason in Political Economy*, London: Taylor and Francis.

Schaefer, H. B. and Singh, R. (2022), 'Property of the Social Media Data', in K. Basu and A. Mishra (eds), *Law and Economic Development: Behavioral and Moral Foundations of a Changing World*, New York: Palgrave Macmillan.

Schlicht, E. (1998), *On Custom in the Economy*, Oxford: Oxford University Press.

Scriven, M. (1951), 'Paradoxical Announcements', *Mind*, 60, 239: 403–7.

Sen, A. (1973), *On Economic Inequality*, Oxford: Oxford University Press.

Sen, A. (1996), 'Hume's Law and Hare's Rule', *Philosophy*, 41(155): 75–9.

Sen, A. (2006), *Identity and Violence*, New York: W. W. Norton & Co.

Sen, A. (2018), *Collective Choice and Social Welfare, An Expanded Edition*, Cambridge, MA: Harvard University Press.

Sethi, R. and Weibull, J. (2016), 'What Is Nash Equilibrium?', *Notices of the AMS*, 63, 5: 526–8.

Shaw, B. (1903), *Man and Superman*, (later edition) London: Penguin Random House.

Sherman, R. (2019), *Uneasy Street: The Anxieties of Affluence*, Princeton: Princeton University Press.

Singh, J. (2006), 'Separation of Powers and the Erosion of the "Right to Property" in India', *Constitutional Political Economy*, 17, 4: 303–24.

Skyrms, B. (2004), *The Stag Hunt and the Evolution of the Social Structure*, Cambridge: Cambridge University Press.

Smith, A. (1776), *An Inquiry into the Nature and Causes of the Wealth of Nations*, London: Strahan and Cadell.

Sri Aurobindo (1997), *Tales of Prison Life*, tr. Sisir Kumar Ghosh, Puducherry: Sri Aurobindo Ashram Publication [Bengali orig. *Karakahini* (1910)].

Stiglitz, J. E. (1975), 'The Theory of Screening, Education, and Distribution of Income', *American Economic Review*, 65: 283–300.

Stiglitz, J. E. (2012), *The Price of Inequality: How Today's Divided Society Endangers Our Future*, New York: W. W. Norton & Co.

Strogatz, S. (2012), 'Friends You Can Count On', *New York Times*, Opinionator, 17 September.

Subramanian, S. (2006), *Measurement of Inequality and Poverty*, New Delhi: Oxford University Press.

Sugden, R. (1995), 'A Theory of Focal Points', *Economic Journal*, 105, 430: 533–50.

Sunstein, C. (1996), 'Social Norms and Social Rules', *Columbia Law Review*, 96, 4: 903–68.

Tait, K. (1977), *My Father, Bertrand Russell*, London: Victor Gollancz Ltd.

Tännsjö, T. (2007), 'The Myth of Innocence: On Collective Responsibility and Collective Punishment', *Philosophical Papers*, 36, 2: 295–314.

Thrasher, J. and Vallier, K. (2013), 'The Fragility of Consensus: Public Reason, Diversity and Stability', *European Journal of Philosophy*, 23, 4: 933–54.

Tullock, G. (1971), 'The Paradox of Revolution', *Public Choice*, 11: 89–99.

Ugander, J., Karrer, B., Backstrom, L. and Marlow, C. A. (2011), 'The Anatomy of the Facebook Social Graph', *ArXiv* abs/1111.4503.

Vanderschraaf, P. (2006), 'War or Peace? A Dynamical Analysis of Anarchy', *Economics & Philosophy*, 22, 2: 243–79.

Vanderschraaf, P. (2019), *Strategic Justice: Conventions and Problems of Balancing Divergent Interests*, New York: Oxford University Press.

Vanderschraaf, P. (2021), 'Contractarianisms and Markets', *Journal of Economic Behavior and Organization*, 181: 270–87.

Veblen, T. (1899), *The Theory of the Leisure Class*, New York: Macmillan.

Villanger, E. (2006), 'Company Interests and Foreign Aid Policy: Playing Donors out against each other', *European Economic Review*, 50, 3: 533–45.

Von Neumann, J. and Morgenstern, O. (1944), *Theory of Games and Economic Behavior*, Princeton: Princeton University Press.

Voorneveld, M. (2010), 'The Possibility of Impossible Stairways: Tail Events and Countable Player Sets', *Games and Economic Behavior*, 68, 1: 403–10.

Walras, L. (1874), *Éléments d'économie politique pure*, 4th edn in 1900, Lausanne: Rouge [English tr. by W. Jaffe as *Elements of Pure Economics*, Philadelphia, 1954].

Weingast, B. (1997), 'The Political Foundations of Democracy and the Rule of the Law', *American Political Science Review*, 91, 2: 245–63.

Weithman, P. (2005), 'Deliberative Character', *Journal of Political Philosophy*, 13, 3: 263–83.

Ye, W. and Fan, S. (2020), 'Evolutionary Traveler's Dilemma Game based on Particle Swarm Optimization', *Physica A: Statistical Mechanics and its Applications*, 544: 1–9.

Index

Kaushik Basu is Carl Marks Professor of International Studies and Professor of Economics at Cornell University. He was Chief Economist of the World Bank from 2012 to 2016, and was previously Chief Economic Advisor to the government of India.